# Taxonomy and paleoecology of late Neogene benthic foraminifera from the Caribbean Sea and eastern equatorial Pacific Ocean

LENNART BORNMALM

Bornmalm, L. 1997 02 15: Taxonomy and paleoecology of late Neogene benthic foraminifera from the Caribbean Sea and eastern equatorial Pacific Ocean. *Fossils and Strata*, No. 41, pp. 1–96. Oslo. ISSN 0300-9491. ISBN 82-00-37666-4.

Benthic foraminifera (147 taxa) were investigated from the Caribbean Sea (Deep Sea Drilling Project; DSDP Hole 502A, depth 3,051 m) and eastern equatorial Pacific Ocean (DSDP Hole 503B, 3,672 m) over an interval between the terminal Miocene and the basal Pleistocene (5.5–1.7 Ma). To determine the influence over time of the Central American Isthmus (closure of the Isthmus of Panama occurring some time between 3.5 and 3.0 Ma) on the benthic fauna, the composition and diversity of the fauna as well as many physico-chemical variables were considered. The latter include changes in coarse-fraction (>63 μm), calcite dissolution (estimated from the degree of fragmentation of planktic foraminifera), accumulation rates of benthic foraminifera (BFAR, used as an index of the flux of organic matter to the sea floor), $CaCO_3$ content, and stable isotopes in planktic and benthic foraminifera. In the Caribbean Hole 502A, the $\delta^{18}O$ of planktic foraminifera increased at about 4.2 Ma relative to the benthic values and also relative to the planktic and benthic values in the Pacific Hole 503. This change may reflect increasing surface-water salinity in the Caribbean Sea as a result of restricted surface-water exchange between the Atlantic and Pacific caused by the emergent Panama Isthmus. Most of the physico-chemical variables in Hole 502A have similar trends to those in the Pacific Hole 503 between late Miocene and 3.85 Ma. After that, Hole 503 continued to show a typical equatorial Pacific character, whereas Hole 502A was affected by local tectonics and bottom-water exchange between Caribbean Sea and the Atlantic. The increased BFAR at 3.85 Ma in Hole 502A may indicate increased productivity of the surface water in the Caribbean Sea, but it could also be a result of a decrease in dissolution and intensified ventilation of the bottom water. The increased $\delta^{18}O$ and $\delta^{13}C$ values in the benthic foraminifer record, increased coarse-fraction $CaCO_3$ content, as well as decreased fragmentation of planktic foraminifera are probably related to greater inflow of Upper North Atlantic Deep Water (UNADW) to the Caribbean Sea since 3.85 Ma, most likely initiated by increased northward transport of warm, high-salinity waters to high latitudes via the Gulf Stream, which in turn was caused by progressive uplift of the Central American land bridge. The fluctuations in the physico-chemical parameters in Hole 503 are overall larger than in Hole 502A, which probably is a result of dissolution signals caused by vertical oscillations of the lysocline in Hole 503 because of the deeper location of this site. The increased planktic and benthic foraminifer oxygen isotope values from 3.2 Ma in Hole 503 probably reflect oceanographic and climatic changes in the Antarctic area from this time. Q-mode principal component analysis on benthic foraminifer abundances (accumulation rates of selected species) distinguished two major faunal groups in both Hole 502A and Hole 503. *Nuttallides umbonifera* is the most abundant species and makes up one group, while the other group is made up of both *Cibicidoides wuellerstorfi* and *Oridorsalis umbonatus* in Hole 502A, and only *O. umbonatus* in Hole 503. In Hole 502A *C. wuellerstorfi* and *O. umbonatus* are abundant in the interval between 4.2 and 3.7 Ma. From about 3.7 Ma *N. umbonifera* increased and became the most abundant benthic species. In Hole 503 *N. umbonifera* was the most abundant species between 4.8 and 2.1 Ma, except for a few intervals where *O. umbonifera* was more frequent. These changes in the abundance of *N. umbonifera* may be due to (1) changes in volume of different water masses, (2) changes in productivity, and/or (3) a transitory shift in an environmental preference of *N. umbonifera*. However, in both Holes 502A and 503 there is no significant correlation among dissolution, $CaCO_3$ content, BFAR, and the variation in the abundance of *N. umbonifera*. □*Benthic foraminifera, biostratigraphy, paleoecology, paleoceanography, late Neogene, Isthmus of Panama, Deep Sea Drilling Project, DSDP Sites 502 and 503, Caribbean Sea, Colombia Basin, eastern equatorial Pacific Ocean, Guatemala Basin.*

*Lennart Bornmalm, Department of Marine Geology, University of Göteborg, P.O. Box 7064, S-402 32 Göteborg, Sweden; 25th February, 1995.*

T0256441

# Contents

# Introduction

Benthic foraminifera are found in most marine environments and may form more than 50% of eukaryotic biomass in the deep sea (Gooday *et al.* 1992). They are one of the most important components of modern and ancient benthic communities and the only organisms that live in large enough numbers in the nutrient-starved ocean-floor environment to be statistically represented in core samples obtained within the Deep Sea Drilling Project (DSDP) and Ocean Drilling Program (ODP) (Thomas 1992). Modern distribution patterns have been used to make inferences about the ecological preferences of fossil assemblages.

Many earlier investigators concluded that there was a distinct relationship between various benthic foraminifer species and/or assemblages and bottom-water masses (Streeter 1973; Schnitker 1974, 1979, 1980; Lohmann 1978; Corliss 1979; Douglas & Woodruff 1981; Hodell *et al.* 1983, 1985; Mead 1985; Hermelin 1986, 1989; Murray 1991, among others). However, the use of benthic foraminifer abundances as unequivocal indicators of the physico-chemical character of the bottom water masses has been doubted in a number of studies (e.g., Corliss 1985; Corliss *et al.*, 1986; Thomas & Vincent 1987, 1988; Linke & Lutze 1993; Gooday 1988, 1993; Mackensen *et al.* 1993; Schnitker 1994). Lutze & Coulbourn (1984) suggested that the distribution of benthic foraminifera may be controlled by the supply of organic matter to the sea floor. More recent studies have shown that a variety of perturbations, including seasonal phenomena, influence the ecology of benthic foraminifer species and populations in at least some deep-sea areas (Tyler 1988; Grassle & Morse-Porteous 1987; Lambshead & Gooday 1990). Fluctuations in food supply are considered to be of particular importance (Jumars & Wheatcroft 1989; Grassle 1989). Therefore, the large influence of nutrient-influx on the deep-sea benthic foraminifer faunas should be kept in mind while assessing the influence of deep-sea circulation on faunal patterns (Gooday & Lambshead 1989; Gooday & Turley 1990; Lambshead & Gooday 1990; Thomas 1992). Futhermore, changes in oxygen concentrations at the sediment–water interface also plays a major role in controlling benthic foraminifer assemblages and morphologic characteristics; such as changes in size, wall thickness, and porosity of foraminifera (Kaiho 1994).

Major tectonic changes, both in the deep-sea and on the continents, have caused changes of the ocean's deep- and shallow-water circulation. One such significant event, which may have affected the benthic foraminifer faunal composition, was apparently the closure of the Central American Seaway, resulting from the emergence of the Panama land-bridge in the middle Pliocene. Previous studies have suggested that the final stage in the isolation of the tropical Atlantic and Pacific oceans occurred between late Miocene and middle Pliocene times (Whitmore & Stewart 1965; Woodring 1966; Kaneps 1970; Emiliani *et al.* 1972; Parker 1973; Casey *et al.* 1975; Rosen 1975; Keigwin 1976). The closure was a gradual process, beginning with a termination of the bottom-water contact in pre-Pliocene times (Gardner 1982). The surface-water connection ceased in middle Pliocene times. Saito (1976) placed the closure of the Isthmus of Panama in the middle Pliocene (about 3.5 Ma). On the other hand, Keigwin (1978) suggested that the closure occurred some half million years later (3.1 Ma).

Gardner (1982) showed that the concentration of calcium carbonate in bulk sediment was approximately the same in Sites 502 and 503 in the interval between 7.5 and 4.0 Ma. Thereafter, the two sites changed with respect to the calcium carbonate concentation: Site 503 continued to reflect the carbonate record of the equatorial Pacific Ocean, whereas tectonic events influenced the $CaCO_3$ content at Site 502.

Keigwin (1982a–c) reported that the circulation patterns in the Caribbean and eastern Pacific developed at about 3.0 Ma, based on planktic foraminifer assemblage changes and analysis of stable isotopes of planktic and benthic foraminifera. Several later studies (Hodell *et al.* 1983, 1985; Brunner 1984; Backman *et al.* 1986) support the inference that the Isthmus of Panama was closed between 3.2 and 2.9 Ma. The evolutionary divergence of eastern Pacific and Caribbean near-shore microfossil organisms suggests a closure of the Isthmus of Panama at approximately 3.5 Ma (Coates *et al.* 1992).

Recently, Vermeij (1993) and Knowlton *et al.* (1993) reported that the gene flow of tropical American mollusks was disrupted between the Caribbean and Pacific by environmental changes several million years before the date for the completion of the landbridge. In addition, Jackson *et al.* (1993) and Allmon *et al.* (1993) suggested that the gradual closure of the Panamanian seaway and the resulting environmental changes stimulated an increase in Caribbean and western Atlantic molluscan diversity rather than causing the mass extinction hypothesized previously on the basis of inadequate data.

Although many studies of deep-sea benthic foraminifera have been carried out in the adjacent Gulf of Mexico (e.g., Phleger 1951; Pflum & Frerichs 1976; Poag 1981, 1984; Gary 1985; Culver 1988; Denne & Sen Gupta 1988, 1991; Gary *et al.* 1989; Corliss & Fois 1990), relatively few studies have been devoted to the Caribbean Sea. Gaby & Sen Gupta (1985), Sen Gupta (1988), and Galluzzo *et al.* (1990) studied the relationship between the bottom-water masses and the benthic foraminifer faunas in the eastern Caribbean Sea.

Many DSDP and ODP sites have been drilled in the Pacific Ocean, but only a few studies have been carried out on Neogene deep-sea benthic foraminifera from the central and eastern part of the Pacific (central North Pacific: Douglas 1973; northeastern Pacific: Ingle 1973; Boltovskoy 1981; eastern margin of the Nazca Plate: Resig 1976, 1981; middle America Slope: McDougall 1985; central equatorial Pacific: Thomas 1985; central equatorial Pacific Ocean: Woodruff 1985; Boersma 1986; Kurihara & Kennett 1986, 1988; eastern Pacific off Peru: Resig 1990; western Pacific: Woodruff & Douglas 1981; Hermelin 1989).

The primary objective of this study was to analyze changes in absolute abundance of deep-sea benthic foraminifer species and diversity of the fauna from the terminal Miocene to the basal Pleistocene (5.5–1.7 Ma) of the Deep Sea Drilling Project (DSDP) Leg 68 Holes 502A and 503, located in the Caribbean Sea and eastern equatorial Pacific Ocean, respectively (Fig. 1). The second objective was to analyze fluctuations in the coarse fraction of the sediment (>63 µm), calcite dissolution based on the degree of fragmentation of planktic foraminifera, flux of organic matter to the sea floor based on accumulation rates of benthic foraminifera (BFAR) (Herguera & Berger 1991; Herguera 1992, 1994; Berger *et al.* 1994), calcium carbonate content, and stable isotopes in planktic and benthic foraminifera. The third objective was to analyze the taxonomy and to illustrate the late Neogene benthic foraminifera from this area using scanning electron photomicrography.

## Study area

The Colombia Basin is a large, deep basin in the western Caribbean Sea. Its depth exceeds 4,000 m, and its connection with the North Atlantic Ocean is over a sill at a depth of about 1,650 m at the Windward Passage between Cuba and Haiti (Sverdrup *et al.* 1942; Wüst 1964; Gordon 1966). The water mass of the western Caribbean Sea (deeper than 50 m) can be divided into four layers, based on dissolved oxygen, temperature, and salinity characteristics (the details have been summarized by Wüst 1964). The Subtropical Underwater (50–200 m) is associated with the warm-water sphere of the ocean and is separated from the lower, cold-water sphere by a layer of low oxygen content (below 3.0 ml/l) from 200 to 600 m. The Subantarctic Intermediate Water (SAIW; found from 700 to 850 m with temperatures of 5°–7°C and a salinity of 34.85‰), and the North Atlantic Deep Water (NADW; situated between 1,800 and 2,500 m with a temperature of 4.1°C and a salinity of 34.98‰) are cold-water layers. The waters between the different layers are made up of mixtures of these water layers, because of the normal vertical mixing by turbulence. According to deMenocal *et al.* (1992) and Raymo *et al.* (1992), who divide NADW into Upper North Atlantic Deep Water (UNADW) and Lower North Atlantic Deep Water (LNADW), the bottom-water mass in the Caribbean Sea contains UNADW. The hydrographic data of the Atlantic Intermediate Water (AIW) flowing over the Windward Passage sill today is mostly (85%) UNADW (temperature 3.8°C; salinity 35‰). The Columbia basin is filled homogeneously from its deepest depth (4,500–5,500 m) to about 1,800 m with this water (Wüst 1964; deMenocal *et al.* 1992). Core top $\delta^{13}C$ values of benthic foraminifera in the Colombia Basin are about 1.0‰ (Keigwin 1982c), which is also consistent with an UNADW source (Oppo & Fairbanks 1987; deMenocal *et al.* 1992). The main outflow for the deep Caribbean water is to the Gulf of Mexico via the Yucatan Passage, which has a sill depth of about 2,000 m (Metcalf 1976).

The Guatemala Basin is located in the eastern equatorial Pacific Ocean, west of Central America. The depth exceeds 4,000 m, and the basin borders the Galapagos Ridge in the south and the central ocean ridge to the west. The subsurface water mass in the Guatemala Basin (below the permanent thermocline at about 50 m) is characterised by the Pacific Equatorial Water. This water mass has its greatest north–south extension along the American coast, where it is present between latitudes 18°S and 20°N (Sverdrup *et al.* 1942). It is probably formed off the coast of South America by gradual transformation of the Subantarctic Water (Sverdrup *et al.* 1942). According to Sverdrup *et al.* (1942) the Pacific Equatorial Water mass is characterized by a nearly straight temperature–salinity correlation with a temperature between 15°C and 8°C and a salinity between 35.15‰ and 34.50‰ above 800 m. At

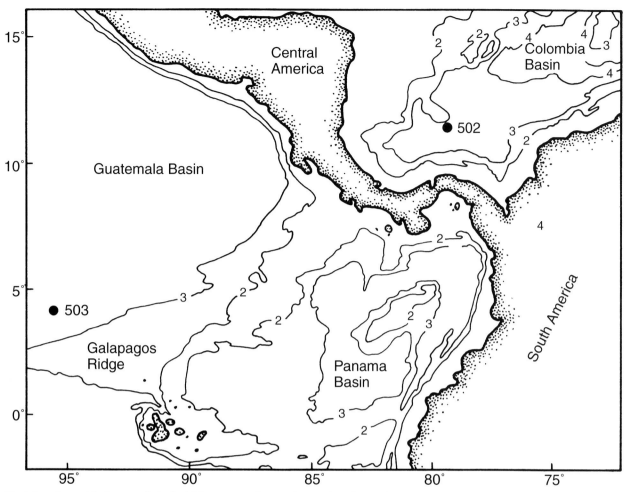

*Fig. 1.* Location and bathymetry (in km) of DSDP Leg 68 Sites 502 (3,051 m) in the Caribbean Sea and 503 in the eastern equatorial Pacific Ocean (3,672 m).

a depth of 800 m, where the temperature is about 5.5°C, a salinity minimum exists, in which the minimum values lie between 34.50‰ and 34.58‰. Below 1,000 m, the temperature and salinity exhibit a nearly straight line, with the temperature decreasing and the salinity increasing toward the bottom, where the corresponding values are 1.3°C and 34.70‰. The presence of strongly dissolved faunas below 3,000 m water depth off Central America has led McDougall (1985) to suggest that the corrosive Antarctic Bottom Water (AABW) has been present there since the Miocene.

## Material and methods

Hole 502A (11°29.46'N and 79°22.74'W) and Hole 503B (4°03.02'N, 95°38.32'W) were drilled at water depths of 3,051 m and 3,672 m, respectively (Prell *et al.* 1982) (Fig. 1). Sediments at Site 502 are composed mostly of nannofossils, with foraminifera as a minor component, and are classified as foraminifer-bearing nannofossil marl. At Site 503, sediments are composed of silica-bearing nannofossil marl, calcareous siliceous ooze, and siliceous nannofossil ooze. In Hole 502A, the coarse-fraction component (>63 μm) consists of planktic foraminifera, whereas the fine fraction is principally composed of nannofossils and juvenile foraminifera (Prell *et al.* 1982). The sequences range in age from the late Miocene to Recent (approximately the last 8 Ma). The recovery was good (82.2% in Hole 502A and 83.5% in Hole 503B) and core deformation was minimal in most intervals (Prell *et al.* 1982). Eighty samples were studied from Hole 502A, from depths between 41.62 and 162.42 m bsf (Table 1). Eighty-nine samples were obtained from Hole 503B (29.31–110.46 m) and six samples from Hole 503A (99.93–138.91 m bsf; Table 2). Samples from Hole 503A were used to expand the stratigraphic interval down to the Miocene–Pliocene boundary. Because of fluctuating abundance due to poor preservation of benthic foraminifera in Hole 503A–B, only 64 samples were used for faunal analysis,

*Table 1.* Depths and estimated ages, total weights of samples, percentages of coarse fraction (>63 mm), fragmentation in planktic foraminifera, BFAR (accumulation rates of benthic foraminifera), and O and C isotope values of *Cibicidoides wuellerstorfi* in samples from Hole 502A (where available).

| Core | Section | Interval (cm) | Depth (m) | Age (Ma) | Weight of sediment (gram) | Coarse fraction >63 μm (%) | Fragmentation in planktic foraminifera (%) | BFAR (No./ (cm$^2$·ka)) | δ$^{18}$O in benthic fora-minifera (‰) | δ$^{13}$C in benthic fora-minifera (‰) |
|---|---|---|---|---|---|---|---|---|---|---|
| 11 | CC | 11–13 | 41.62 | 1.71 | 20.70 | 20.0 | 11.7 | 61 | 2.79 | 0.45 |
| 12 | 1 | 96–99 | 46.87 | 1.88 | 15.58 | 50.7 | 51.8 | l38 | 2.53 | 1.02 |
| 12 | 2 | 47–50 | 47.76 | 1.91 | 17.96 | 46.2 | 27.9 | 139 | 2.91 | 0.55 |
| 12 | 3 | 0–3 | 48.80 | 1.94 | 13.40 | 14.8 | 8.2 | 117 | 2.60 | 0.48 |
| 12 | 3 | 88–90 | 49.67 | 1.97 | 14.52 | 54.6 | 26.7 | 102 | | |
| 13 | 1 | 12–15 | 50.43 | 2.00 | 19.19 | 26.9 | 44.0 | 73 | 2.88 | 0.85 |
| 16 | 1 | 4–6 | 63.55 | 2.44 | 4.43 | 10.6 | 34 3 | 334 | | |
| 16 | 1 | 100–103 | 64.51 | 2.47 | 14.54 | 32.9 | 42.4 | 147 | 3.02 | 0.69 |
| 16 | 2 | 51–53 | 65.51 | 2.50 | 14.71 | 28.4 | 61.3 | 108 | 2.52 | 0.70 |
| 16 | 3 | 0–3 | 66.51 | 2.53 | 17.56 | 24.0 | 67.8 | 111 | 2.36 | 0.70 |
| 17 | 1 | 51–53 | 68.42 | 2.60 | 16.23 | 29.8 | 58.2 | 53 | 2.38 | 0.79 |
| 17 | 2 | 0–3 | 69.43 | 2.63 | 7.01 | 26.1 | 52.9 | 237 | | |
| 17 | 2 | 100–102 | 70.42 | 2.66 | 13.18 | 27.3 | 66.4 | 94 | 2.42 | 0.25 |
| 17 | 3 | 51–53 | 71.42 | 2.70 | 13.03 | 18.9 | 66.4 | 126 | 2.49 | 1.01 |
| 17 | CC | 0–3 | 72.39 | 2.73 | 15.23 | 15.7 | 67.6 | 146 | 2.59 | 0.84 |
| 19 | 1 | 0–3 | 76.72 | 2.87 | 11.91 | 12.1 | 9.55 | 90 | 2.33 | 0.79 |
| 19 | 1 | 98–100 | 77.69 | 2.91 | 13.85 | 17.8 | 57.2 | 115 | 2.45 | 0.77 |
| 19 | 2 | 51–53 | 78.45 | 2.93 | 14.94 | 25.2 | 70.0 | 154 | 2.06 | 0.60 |
| 19 | 3 | 0–3 | 79.47 | 2.97 | 12.83 | 19.3 | 16.9 | 139 | 2.16 | 0.67 |
| 19 | CC | 0–3 | 80.46 | 3.00 | 14.66 | 24.2 | 57.5 | 127 | 2.41 | 0.48 |
| 20 | I | 97–99 | 81.91 | 3.05 | 9.51 | 10.0 | 10.8 | 217 | | |
| 20 | 2 | 55–57 | 83.03 | 3.08 | 12.64 | 19.0 | 48.8 | 106 | | |
| 20 | 3 | 0–3 | 83.99 | 3.10 | 12.55 | 21.1 | 15.7 | 107 | 2.18 | 0.63 |
| 20 | 3 | 97–99 | 84.96 | 3.13 | 12.81 | 18.9 | 10.3 | 105 | 2.40 | 0.64 |
| 21 | l | 55–57 | 86.06 | 3.18 | 8.96 | 16.3 | 58.2 | 89 | | |
| 21 | 2 | 0–3 | 86.98 | 3.21 | 12.10 | 19.6 | 49.4 | 119 | | |
| 21 | 2 | 101–104 | 88.02 | 3.25 | 18.80 | 13.8 | 19.9 | 72 | 2.26 | 0.50 |
| 21 | 3 | 37–39 | 88.84 | 3.28 | 13.91 | 14.8 | 16.8 | 69 | | |
| 21 | 3 | 50–53 | 88.97 | 3.29 | l5.89 | 5.9 | 43.7 | 39 | 1.90 | 0.52 |
| 21 | CC | 0–3 | 89.58 | 3.31 | 16.14 | 12.1 | 41.3 | 132 | 2.14 | 0.70 |
| 22 | 1 | 103–105 | 90.94 | 3.36 | 6.72 | 13.0 | 41.5 | 167 | 2.00 | 0.33 |
| 22 | 2 | 43–45 | 91.85 | 3.40 | 13.95 | 6.9 | 41.2 | 97 | 2.06 | 0.66 |
| 22 | 3 | 0–3 | 92.91 | 3.43 | 15.05 | 5.2 | 44.3 | 126 | 2.22 | 0.55 |
| 22 | 3 | 101–104 | 94.48 | 3.45 | 13.58 | 15.1 | 10.1 | 114 | 2.13 | 0.44 |
| 23 | l | 46–48 | 94.77 | 3.48 | 14.55 | 13.9 | 43.5 | 157 | | |
| 23 | 2 | 105–107 | 96.82 | 3.54 | 6.30 | 17.5 | 41.7 | 306 | | |
| 23 | 3 | 52–54 | 97.80 | 3.56 | 15.06 | 15.4 | 33.5 | 124 | 1.87 | 0.54 |
| 23 | CC | 2–4 | 98.81 | 3.59 | 6.26 | 14.2 | 49.9 | 315 | | |
| 24 | 1 | 102–104 | 99.73 | 3.62 | 7.16 | 9.2 | 44.7 | 204 | | |
| 24 | 2 | 51–53 | 100.73 | 3.65 | 13.70 | 9.6 | 46.3 | 185 | 1.96 | 0.57 |
| 25 | 1 | 52–54 | 103.63 | 3.73 | 12.97 | 7.9 | 53.8 | 62 | 2.03 | 0.48 |
| 25 | 2 | 1–3 | 104.54 | 3.75 | 17.20 | 21.3 | 54.3 | 110 | 2.13 | 0.53 |
| 25 | 2 | 105–108 | 105.58 | 3.78 | 14.89 | 11.6 | 53.8 | 113 | 1.92 | 0.47 |
| 25 | 3 | 50–53 | 106.53 | 3.81 | 1.52 | 22.4 | 39.8 | 379 | | |
| 26 | 1 | 15–17 | 107.66 | 3.84 | 5.36 | 8.2 | 31.0 | 284 | 2.02 | 0.53 |
| 26 | 1 | 115–117 | 108.66 | 3.87 | 14.93 | 8.5 | 56.4 | 75 | 1.83 | 0.17 |
| 26 | 2 | 57–59 | l09.51 | 3.89 | 15.61 | 12.5 | 67.8 | 45 | 1.93 | 0.37 |
| 26 | 3 | l–3 | 110.47 | 3.92 | 15.83 | 8.0 | 62.2 | 77 | 2.39 | 0.26 |
| 26 | 3 | 102–104 | 111.48 | 3.94 | 15.85 | 8.6 | 61.1 | 73 | 2.14 | 0.33 |
| 27 | 1 | 50–53 | 112.42 | 3.97 | 15.41 | 15.0 | 47.9 | 114 | | |
| 27 | 2 | 1–3 | 113.42 | 4.00 | 15.96 | 7.8 | 49.9 | 68 | l.93 | 0.26 |
| 27 | 2 | 104–106 | 114.45 | 4.03 | 14.95 | 4.6 | 62.5 | 59 | 1.97 | 0.37 |
| 28 | 1 | 21–23 | 116.52 | 4.08 | 15.67 | 3.7 | 74.4 | 39.31 | 2.11 | 0.48 |
| 28 | 1 | 118–120 | 117.79 | 4.13 | 12.58 | 2.7 | 62.4 | 48.50 | 1.90 | 0.48 |
| 28 | 2 | 68–70 | 118.33 | 4.15 | 15.36 | 3.3 | 13.6 | 70.39 | 1.99 | 0.25 |
| 28 | 3 | 18–20 | 119.33 | 4.19 | 15.07 | 5.4 | 69.8 | 44.97 | | |
| 29 | 1 | 96–98 | 120.27 | 4.23 | 12.78 | 3 7 | 74.9 | 34.61 | | |
| 29 | 2 | 45–48 | 121.26 | 4.27 | 14.80 | 3.9 | 74.5 | 38.98 | 2.12 | 0.13 |
| 29 | 3 | 0–3 | 122.32 | 4.31 | 13.00 | 3.7 | 29.1 | 140.84 | 2.20 | 0.18 |
| 30 | 1 | 51–54 | 123.32 | 4.34 | 12.96 | 2.1 | 60.9 | 74.76 | 1.85 | 0.16 |
| 30 | 2 | 0–3 | 124.29 | 4.38 | 13.55 | 7.2 | 79.5 | 77.53 | | |
| 30 | 2 | 105–107 | 125.35 | 4.42 | 14.56 | 5.2 | 72.3 | 119.45 | 2.27 | 0.35 |

*Table 1* (continued).

| Core | Section | Interval (cm) | Depth (m) | Age (Ma) | Weight of sediment (gram) | Coarse fraction >63 μm (%) | Fragmentation in planktic foraminifera (%) | BFAR (No./ (cm²·ka)) | δ¹⁸O in benthic fora-minifera (‰) | δ¹³C in benthic fora-minifera (‰) |
|---|---|---|---|---|---|---|---|---|---|---|
| 30 | 3 | 0–3 | 125.80 | 4.43 | 16.56 | 3.6 | 33.7 | 64.43 | | |
| 32 | 1 | 5–8 | 128.36 | 4.52 | 15.74 | 7.1 | 85.7 | 70.56 | 1.87 | 0.19 |
| 32 | 1 | 103–106 | 129.34 | 4.56 | 14.36 | 3.4 | 76.3 | 78.37 | 1.99 | 0.18 |
| 32 | 2 | 47–49 | 130.30 | 4.59 | 5.24 | 17.9 | 34.8 | 88.20 | | |
| 32 | 3 | 4–6 | 131.38 | 4.63 | 16.43 | 10.2 | 83.1 | 46.03 | | |
| 33 | 1 | 96–98 | 132.27 | 4.66 | 15.08 | 4.8 | 56.4 | 57.96 | 1.92 | 0.20 |
| 33 | 2 | 48–50 | 133.26 | 4.70 | 15.37 | 3.7 | 71.0 | 72.66 | | |
| 33 | 3 | 0–2 | 134.28 | 4.74 | 15.35 | 5.5 | 50.8 | 65.09 | 2.08 | 0.42 |
| 34 | 1 | 102–104 | 135.33 | 4.77 | 15.71 | 6.4 | 76.7 | 44.68 | 1.90 | 0.50 |
| 34 | 2 | 50–52 | 136.31 | 4.82 | 4.10 | 12.2 | 63.5 | 303.33 | | |
| 34 | 2 | 142–145 | 137.23 | 4.87 | 15.34 | 8.8 | 77.3 | 51.59 | | |
| 35 | 1 | 103–105 | 138.34 | 4.92 | 16.49 | 14.4 | 62.4 | 72.89 | 1.89 | 0.39 |
| 35 | 2 | 46–48 | 138.82 | 4.95 | 14.79 | 9.4 | 71.2 | 41.83 | 1.83 | 0.42 |
| 35 | 3 | 1–3 | 139.87 | 5.00 | 14.60 | 10.0 | 67.8 | 88.42 | 1.92 | 0.39 |
| 37 | 2 | 8–10 | 144.03 | 5.21 | 14.77 | 9.1 | 64.1 | 64.44 | | |
| 40 | 2 | 1–3 | 151.70 | 5.37 | 14.34 | 19.5 | 55.2 | 117.04 | 2.37 | 0.32 |
| 42 | 2 | 48–50 | 157.79 | 5.46 | 14.60 | 9.3 | 30.1 | 93.30 | 2.19 | 0.36 |
| 44 | 2 | 30–33 | 162.42 | 5.54 | 14.41 | 8.7 | 29.8 | 40.04 | 2.12 | 0.46 |

*Table 2.* Depths and estimated ages, total weights of the samples, percentages of the coarse fraction (>63 μm), and fragmentation in planktic foraminifera and BFAR (accumulation rates of benthic foraminifera) in samples from Hole 503.

| Hole | Core | Section | Interval (cm) | Depth (m) | Age (Ma) | Weight of sediment (g) | Coarse fraction >63 μm (%) | Fragmentation in plank-tic foraminifera (%) | BFAR (No./(cm²·ka) |
|---|---|---|---|---|---|---|---|---|---|
| 503B | 8 | 1 | 10–12 | 29.31 | 1.65 | 11.30 | 33.80 | 70.0 | |
| | 8 | 1 | 61–63 | 29.82 | 1.66 | 4.10 | 4.60 | 54.9 | 36 |
| | 8 | 2 | 10–12 | 30.71 | 1.70 | 3.29 | 4.00 | 52.4 | |
| | 8 | 2 | 115–117 | 31.76 | 1.75 | 2.26 | 3.50 | 70.5 | |
| | 8 | 3 | 66–68 | 32.78 | 1.79 | 2.67 | 8.20 | 68.2 | |
| | 8 | 3 | 116–118 | 33.20 | 1.81 | 5.18 | 6.00 | 76.8 | 67 |
| | 10 | 1 | 12–14 | 38.11 | 2.01 | 3.50 | 4.90 | 59.5 | 60 |
| | 10 | 1 | 112–114 | 39.13 | 2.05 | 6.50 | 30.30 | 65.1 | 34 |
| | 10 | 2 | 10–12 | 39.58 | 2.07 | 4.48 | 11.60 | 62.2 | 61 |
| | 10 | 2 | 60–62 | 40.09 | 2.09 | 4.93 | 10.30 | 60.6 | 39 |
| | 10 | 3 | 10–13 | 41.08 | 2.14 | 2.87 | 4.50 | 51.2 | 82 |
| | 11 | 1 | 15–17 | 42.56 | 2.20 | 11.45 | 64.60 | | |
| | 11 | 2 | 11–13 | 43.79 | 2.25 | 3.01 | 7.60 | 68.1 | 75 |
| | 11 | 2 | 61–63 | 44.29 | 2.27 | 3.03 | 27.10 | 76.0 | 131 |
| | 11 | 2 | 112–114 | 44.80 | 2.29 | 4.30 | 21.90 | 60.5 | 95 |
| | 11 | 3 | 62–64 | 45.80 | 2.33 | 3.90 | 23.10 | 48.4 | 59 |
| | 17 | 1 | 7–9 | 46.85 | 2.38 | 3.28 | 10.40 | 68.4 | 46 |
| | 12 | 1 | 57–59 | 47.38 | 2.40 | 2.50 | 10.80 | 56.1 | |
| | 12 | 1 | 106–108 | 47.87 | 2.42 | 3.02 | 5.60 | 47.1 | |
| | 12 | 2 | 57–59 | 48.88 | 2.46 | 3.82 | 6.50 | 55.1 | 95 |
| | 12 | 3 | 7–9 | 49.88 | 2.53 | 7.13 | 54.00 | | |
| | 13 | 1 | 22–24 | 51.43 | 2.57 | 4.27 | 4.20 | 32.0 | 55 |
| | 13 | 1 | 75–77 | 51.96 | 2.59 | 4.64 | 5.00 | 65.3 | 85 |
| | 13 | 1 | 126–128 | 52.47 | 2.61 | 5.24 | 10.70 | 60.5 | 71 |
| | 13 | 2 | 81–84 | 52.94 | 2.63 | 4.25 | 9.60 | 43.3 | 47 |
| | 13 | 3 | 37–39 | 54.53 | 2.69 | 4.58 | 3.30 | 85.2 | 86 |
| | 14 | 1 | 28–30 | 55.89 | 2.75 | 10.63 | 75.80 | | |
| | 14 | 1 | 78–80 | 56.39 | 2.78 | 3.59 | 6.70 | 84.1 | 94 |
| | 14 | 1 | 128–130 | 58.89 | 2.79 | 4.15 | 10.90 | 69.2 | 53 |
| | 14 | 2 | 128–130 | 58.00 | 2.84 | 4.86 | 13.00 | 60.9 | 52 |
| | 14 | 3 | 28–30 | 58.51 | 2.87 | 4.37 | 11.40 | 60.4 | 48 |
| | 14 | 3 | 77–79 | 59.00 | 2.89 | 4.39 | 10.90 | 63.8 | 81 |
| | 15 | 1 | 57–60 | 60.58 | 2.94 | 2.28 | 17.10 | 80.3 | |
| | 15 | 2 | 61–63 | 61.77 | 2.99 | 2.99 | 5.40 | 82.0 | 107 |
| | 15 | 3 | 11–13 | 62.77 | 3.03 | 3.01 | 3.00 | 74.3 | 77 |
| | 15 | 3 | 61–63 | 62.27 | 3.07 | 3.56 | 11.50 | 74.4 | 51 |

*Table 2* (continued).

| Hole | Core | Section | Interval (cm) | Depth (m) | Age (Ma) | Weight of sediment (g) | Coarse fraction >63 μm (%) | Fragmentation in plank- tic foraminifera (%) | BFAR (No./(cm²·ka) |
|------|------|---------|---------------|-----------|----------|------------------------|----------------------------|----------------------------------------------|--------------------|
|  | 15 | CC | 6–8 | 63.87 | 3.08 | 3.05 | 14.10 | 70.9 | |
|  | 16 | 1 | 44–46 | 64.85 | 3.11 | 4.30 | 4.70 | 80.7 | 41 |
|  | 16 | 1 | 96–98 | 65.37 | 3.13 | 5.46 | 9.80 | 62.2 | 53 |
|  | 16 | 1 | 146–148 | 65.87 | 3.14 | 4.65 | 13.10 | 45.6 | 29 |
|  | 16 | 2 | 51–53 | 66.26 | 3.16 | 3.46 | 3.50 | 75.8 | |
|  | 16 | 2 | 100–102 | 60.75 | 3.17 | 4.40 | 18.00 | 69.7 | 22 |
|  | 16 | 3 | 1–3 | 67.20 | 3.19 | 5.02 | 15.90 | 77.5 | 58 |
|  | 16 | 3 | 54–56 | 67.79 | 3.20 | 5.05 | l6.60 | 76.6 | 35 |
|  | 16 | 3 | 102–104 | 68.27 | 3.22 | 4.70 | 19.10 | 74.1 | 43 |
|  | 17 | 1 | 14–16 | 65.95 | 3.28 | 4.43 | 12.00 | 73.3 | 93 |
|  | 17 | 1 | 62–64 | 69.43 | 3.29 | 3.38 | 12.10 | 66.5 | 70 |
|  | 17 | 1 | 118–120 | 69.99 | 3.31 | 3.99 | 11.30 | 67.8 | 103 |
|  | 17 | 2 | 22–24 | 70.44 | 3.33 | 4.37 | 7.80 | 68.7 | 40 |
|  | 17 | 2 | 74–76 | 70.96 | 3.34 | 3.18 | 4.40 | 53.5 | 82 |
|  | 17 | 2 | 126–128 | 71.48 | 3.36 | 3.41 | 11.10 | 70.3 | 97 |
|  | 17 | 3 | 31–33 | 72.03 | 3.38 | 2.91 | 8.90 | 63.6 | 71 |
|  | 17 | 3 | 85–87 | 72.61 | 3.40 | 3.39 | 9.40 | 72.8 | 76 |
|  | 18 | 1 | 13–15 | 73.34 | 3.42 | 4.98 | 8.20 | 76.3 | 119 |
|  | 18 | 1 | 61–63 | 73.82 | 3.44 | 3.75 | 9.90 | 59.8 | 85 |
|  | 18 | 1 | 111–113 | 74.32 | 3.45 | 3.18 | 9.10 | 57.7 | 110 |
|  | 18 | 2 | 11–13 | 74.82 | 3.46 | 3.43 | 7.60 | 67.1 | 169 |
|  | 18 | 2 | 64–66 | 75.35 | 3.46 | 3.56 | 2.20 | 68.7 | 74 |
|  | 18 | 3 | 14–16 | 76.36 | 3.47 | 4.27 | 3.30 | 77.4 | 60 |
|  | 18 | 3 | 113–115 | 7735 | 3.50 | 3.93 | 3.60 | 88.7 | 80 |
|  | 19 | 1 | 73–75 | 79.34 | 3.53 | 3.46 | 3.80 | 81.7 | |
|  | 19 | 2 | 21–23 | 79.13 | 3.55 | 3.49 | 5.40 | 77.4 | |
|  | 19 | 2 | 128–130 | 50.20 | 3.60 | 3.62 | 12.70 | 73.2 | |
|  | 17 | 3 | 28–30 | 80.70 | 3.60 | 2.82 | 2.50 | 95.2 | |
|  | 20 | 1 | 60–62 | 81.77 | 3.63 | 2.44 | 20.50 | 75.2 | 104 |
|  | 20 | 2 | 62–64 | 82.63 | 3.66 | 1.83 | 8.70 | 55.6 | |
|  | 20 | 2 | 113–115 | 83.14 | 3.68 | 1.66 | 15.70 | 35.8 | |
|  | 20 | 3 | 61–63 | 84.90 | 3.71 | 1.83 | 6.00 | | |
|  | 20 | 3 | 111–113 | 84.59 | 3.72 | 5.05 | 2.40 | 95.8 | |
|  | 21 | 1 | 3–5 | 86.44 | 3.79 | 4.18 | 72.20 | | |
|  | 21 | 1 | 56–58 | 86.97 | 3.79 | 4.93 | 10.50 | 80.9 | 57 |
|  | 21 | 1 | 103–105 | 87.44 | 3.81 | 1.53 | 28.70 | 83.7 | |
|  | 21 | 2 | 55–57 | 88.41 | 3.84 | 3.97 | 14.50 | 74.7 | 70 |
|  | 21 | 3 | 12–14 | 89.48 | 3.87 | 3.10 | 16.70 | 56.3 | 143 |
|  | 21 | 3 | 118–120 | 90.S4 | 3.90 | 2.09 | 10.30 | 50.6 | |
|  | 22 | 1 | 56–58 | 91.37 | 3.92 | 5.46 | 6.60 | 68.6 | 48 |
|  | 22 | 2 | 11–13 | 92.17 | 3.93 | 1.82 | 8.60 | | |
|  | 22 | 2 | 111–113 | 93.17 | 3.95 | 2.66 | 34.20 | 33.3 | 100 |
|  | 22 | 3 | 61–53 | 94.17 | 3.97 | 4.07 | 23.80 | 32.1 | |
|  | 22 | 3 | 111–113 | 94.67 | 3.98 | 4.88 | 58.50 | 53.0 | |
|  | 24 | 1 | 61–63 | 100.22 | 4.10 | 4.25 | 3.50 | 39.2 | |
|  | 24 | 2 | 11–13 | 100.77 | 4.11 | 5.39 | 9.50 | 61.0 | 83 |
|  | 24 | 2 | 66–68 | 101.32 | 4.12 | 3.53 | 9.l0 | 62.3 | |
|  | 25 | 2 | 22–24 | 105.72 | 4.19 | 3.05 | 8.40 | 63.5 | |
|  | 25 | 2 | 133–135 | 106.38 | 4.21 | 3.80 | 6.60 | 55.3 | 64 |
|  | 26 | 1 | 8–10 | 108.49 | 4.25 | 2.94 | 7.50 | 63.3 | 113 |
|  | 26 | 1 | 108–110 | 109.49 | 4.27 | 4.00 | 39.10 | 62.3 | 71 |
|  | 26 | 2 | 67–69 | 109.93 | 4.28 | 3.01 | 6.50 | 41.6 | 144 |
|  | 26 | 2 | 120–122 | 110.46 | 4.30 | 8.77 | 18.30 | 69.2 | 45 |
| 503A | 24 | 1 | 132–134 | 99.93 | 4.18 | 4.17 | 14.50 | | |
|  | 27 | 1 | 22–25 | 112.04 | 4.51 | 4.75 | 10.30 | 68.2 | |
|  | 29 | 2 | 61–63 | 112.72 | 4.80 | 7.04 | 6.70 | 55.0 | 34 |
|  | 30 | 2 | 41–43 | 126.75 | 4.90 | 7.21 | 1.90 | 53.3 | 129 |
|  | 31 | 3 | 71–73 | 133.07 | 5.07 | 5.67 | 5.30 | 40.7 | 89 |
|  | 33 | 1 | 70–72 | 138.91 | 5.23 | 3.80 | 10.80 | 77.6 | |

*Table 3.* Ages and depths in sites of magnetostratographic and biostratigraphic datum levels used here in the development of a chronology for Hole 502 and Hole 503. The magnetostratigraphy for Hole 502A, 502 B and Hole 503 is from Kent & Spariosu (1982). Depth of the first appearance datum (FAD) of the planktic foraminifer *Globorotalia tumida* is from Keigwin (1982a), and depth of the magnetic Chron-6 $\delta^{13}C$ shift is from Keigwin (1982c). Depth of the magnetostratigraphic boundaries and the carbon isotope shift listed here are means of upper and lower depth limits given in these articles. Ages are calibrated to the timescale of Berggren *et al.* (1985).

| Datum level | | Age (Ma) | Hole 502A Depth (m) | Hole 502B Depth (m) | Hole 503 Depth (m) |
|---|---|---|---|---|---|
| Olduvai | Top | 1.66 | 39.70 | – | – |
| | Bottom | 1.88 | 46.85 | 42.45 | – |
| Matuyama/Gauss boundary | | 2.47 | – | – | 47.95 |
| Kaena | Top | 2.92 | 78.13 | 73.79 | – |
| | Bottom | 2.99 | 80.05 | 75.54 | – |
| Mammoth | Top | 3.08 | 83.55 | 79.15 | – |
| | Bottom | – | – | – | – |
| Gauss/Gilbert boundary | | 3.40 | 91.90 | – | – |
| Cochiti | Top | 3.88 | – | – | 88.65 |
| Nunivak | Top | 4.10 | 117.10 | – | – |
| | Bottom | 4.24 | 120.40 | – | – |
| Gilbert C2 | Bottom | 4.77 | 135.25 | – | – |
| FAD *G. tumida* | | 5.30 | 145.14 | – | 139.27 |
| Magnetic Chron-6 $\delta^{13}C$ shift | | 6.10 | 197 | – | – |

since the minimum sample size for inclusion into the faunal analysis was set at 50 benthic foraminifera. The combined sequence from Hole 503B and A is hereafter referred to as Hole 503.

To establish a chronology, the paleomagnetic reversals shown in Table 3 (Kent & Spariosu 1982) and the Magnetic Chron-6 carbon shift (about 6.1 Ma; Keigwin 1982c) were used as reference points. The ages were calibrated to the time scale of Berggren *et al.* (1985).

The sedimentation rates were relatively high throughout much of Sites 502 and 503, averaging 3.0 cm/ka at Site 502 and 2.9 cm/ka at Site 503. Interpolations between magnetochronologic datum levels show that the section analyzed in Hole 502A spans the interval between 1.71 and 5.54 Ma (Table 1), whereas that in Hole 503 spans the interval between 1.65 and 5.23 Ma (Table 2). The sampling interval in Hole 502A is 0.1–7.7 m, which corresponds to a sampling resolution of <0.01–0.20 Ma, and between 0.45 and 7.60 m in Hole 503, which corresponds to a sampling resolution of 0.04–0.50 Ma. The average sampling resolution is 0.05 Ma in both holes (1.51 m in Hole 502A and 1.61 m in Hole 503).

Approximately 10 cm³ of each sample was immersed in de-ionized water and placed on a rotary table for about 24 hrs. It was than washed over a 63 μm sieve. The <63 and >63 μm fractions were dried separately. Coarse-fraction

percentages (>63 μm fraction) were determined from the weights of these fractions. The benthic foraminifera from the >125 μm fraction were picked; the number of specimens per sample ranged from 55 to 504 in Hole 502A and from 50 to 195 in Hole 503 (Appendix).

In order to establish dissolution indices, the degree of fragmentation of planktic foraminiferal tests were determined. Although increased fragmentation may be produced during processing the samples in the laboratory, the degree of fragmentation of planktic foraminiferal tests is generally considered to represent an important index of calcite dissolution (Thiede 1973; Thunell 1976; Malmgren 1983). Le & Shackleton (1992) pointed out that this measure, in contrast to other measures of dissolution, such as ratios between dissolution-resistant and dissolution-susceptible species, is essentially independent of ecologic influence. The degree of fragmentation as the relative abundance (percentage) of fragments of planktic foraminifera was computed in relation to whole and fragmented tests of planktic foraminifera. A fragment is here considered to represent a specimen in which less than half of the test remained.

Sedimentation accumulation rates (SAR) were estimated as sedimentation rate (cm/ka) multiplied with the mean density of the sediment, $\delta_w c$ (g/cm³) for a particular interval (Prell *et al.* 1982). Benthic foraminifer accumulation rates (BFAR) were estimated by multiplication of the SAR value with the number of benthic foraminifera per gram sediment (Herguera & Berger 1991; Herguera 1992; Berger & Herguera 1992). The calcium carbonate data used here are from Gardner (1982).

In Hole 502A, analyses of stable isotopes were performed on *Cibicidocides wuellerstorfi* from the size fraction >125 μm. Between 5 and 18 individuals were picked from each sample and cleaned in an ultrasonic bath to remove fine-fraction contamination. All samples were crushed in methanol and roasted under vacuum at 400°C for 30 minutes. $CO_2$ was extracted from the carbonate by reaction with 100% orthophosphoric acid at 50°C. The gaseous samples were purified in two steps with (1) $CO_2$ and (2) frozen ethanol and then analyzed on a Vg Micromass 903E mass spectrometer. Calibration to the PDB standard was achieved through the NBS-20 reference sample. The analytical precision of this method (including $CO_2$ preparation and spectrometric analysis) was 0.04‰ for $\delta^{18}O$ and 0.03‰ for $\delta^{13}C$. All data are reported to PDB by the standard notation. Oxygen and carbon isotope data for planktic foraminifera (the *Globigerinoides trilobus* – *G. quadrilobatus* – *G. sacculifera* group) presented by Keigwin (1982b–c) are used for comparison. The stable isotope data for both planktic and benthic foraminifera in Hole 503 are from Keigwin (1982b–c).

*Q*-mode principal components analysis was used to study faunal changes through the interval. The principal components analysis was based on estimates of accumulation rates (fluxes) of those 12 taxa in Hole 502A and 13 taxa in Hole 503 that showed an accumulation greater than or equal to 1 specimen/(cm$^2$·ka) in at least 50% of the samples. These species were regarded as sufficiently abundant to be reliably used in a multivariate faunal analysis.

Measurements of diversity using the number of species are dependent upon sample size, since larger samples generally contain more species than smaller samples (de Caprariis *et al.* 1976; Douglas 1973; Kempton 1979; Malmgren & Sigaroodi 1985; Parker *et al.* 1984). Therefore, Hurlbert's diversity index was employed ($S[m]$; Hurlbert 1971) to circumvent differences in sample sizes. $S(m)$ denotes the estimated number of species (taxa) expected in a subsample of $m$ specimens taken at random from a sample of $N$ specimens ($m = <N$). In the present study, $m$ was set to 50, since the sample size ranged between 50 and 504 (Tables 4 and 5).

*Acknowledgements.* – This paper is part of my doctoral dissertation. I thank my major advisor Björn Malmgren, Department of Marine Geology, University of Göteborg, for his constant encouragement, excellent guidance and care throughout this investigation. I am also very thankful to Otto Hermelin, Department of Geology, University of Stockholm, for generously providing the material from DSDP Sites 502 and 503. Thanks go to Otto Hermelin, Qingmin Miao, Walter Wheeler, and Kathy Tedesco for comments on an earlier version of the manuscript. Ellen Thomas is gratefully acknowledged for constructive reviews of later version of the manuscript. I am also grateful to Owe Gustafsson for isotopic analysis, and to Gun Karlsson, Ursula Schwarz, Termeh Alavi, and Teodora Andinsson, Department of Marine Geology, University of Göteborg, who provided laboratory assistance. Marjatta Eliason, University of Göteborg, and Amber Taylor, Duke University, are acknowledged for preparing the drawings.

This work was supported by grants from the Royal Swedish Academy of Sciences (Hierta–Retzius' stipendiefond and Th. Nordström's fond för vetenskaplig forskning) and Adlerbertska Forskningsfonden, University of Göteborg.

*Table 4.* Diversity of benthic foraminifera (standardized number of species $S_{50}$) and loadings of the first (PC1) and second (PC2) varimax rotated Q-mode principal components axes in samples from Hole 502A. These account for 90% of the variability in sample space.

| Age (Ma) | Diversity | PC1 | PC2 | Age (Ma) | Diversity | PC1 | PC2 |
|---|---|---|---|---|---|---|---|
| 1.71 | 10.1 | 0.8790 | 0.1494 | 3.65 | 16.0 | 0.6992 | 0.6659 |
| 1.88 | 13.8 | 0.9794 | 0.1468 | 3.73 | 17.1 | 0.9383 | 0.2406 |
| 1.91 | 12.2 | 0.1864 | 0.6881 | 3.75 | 17.5 | 0.0197 | 0.9256 |
| 1.94 | 12.7 | 0.9723 | 0.1698 | 3.78 | 18.3 | 0.0335 | 0.9694 |
| 1.97 | 14.3 | 0.9485 | 0.2525 | 3.84 | 20.1 | 0.1325 | 0.7103 |
| 2.00 | 17.4 | 0.7705 | 0.5842 | 3.87 | 19.2 | 0.0043 | 0.9606 |
| 2.44 | 14.2 | 0.9323 | 0.3310 | 3.89 | 22.7 | 0.0526 | 0.8612 |
| 2.47 | 15.0 | 0.9346 | 0.2838 | 3.92 | 18.3 | 0.2275 | 0.9054 |
| 2.50 | 15.1 | 0.8208 | 0.5071 | 3.94 | 17.3 | 0.1082 | 0.9464 |
| 2.53 | 14.9 | 0.8621 | 0.4500 | 3.97 | 17.9 | 0.8483 | 0.4404 |
| 2.60 | 16.5 | 0.9267 | 0.2994 | 4.00 | 19.5 | 0.0428 | 0.9846 |
| 2 63 | 20.2 | 0.5641 | 0.7663 | 4.03 | 19.7 | 0.3840 | 0.7863 |
| 2.66 | 19.0 | 0.9179 | 0.3043 | 4.08 | 19.4 | 0.2922 | 0.9042 |
| 2.70 | 21.7 | 0.7203 | 0.5107 | 4.13 | 18.1 | 0.6192 | 0.7267 |
| 2.73 | 17.5 | 0.9159 | 0.3532 | 4.15 | 18.8 | 0.7117 | 0.6486 |
| 2 87 | 19.1 | 0.0229 | 0.7977 | 4.19 | 19.0 | 0.3216 | 0.8580 |
| 2.91 | 14.7 | 0.6577 | 0.6486 | 4.23 | 20.0 | 0.5874 | 0.7524 |
| 2.93 | 18.7 | 0.8919 | 0.4082 | 4.27 | 21.6 | 0.8860 | 0.4069 |
| 2.97 | 13.4 | 0.9888 | 0.0833 | 4.31 | 16.3 | 0.9296 | 0.3054 |
| 3.00 | 18.6 | 0.8895 | 0.3458 | 4.34 | 17.7 | 0.8185 | 0.5134 |
| 3.05 | 11.0 | 0.9949 | 0.0797 | 4.38 | 18.1 | 0.6628 | 0.6746 |
| 3.08 | 16.0 | 0.9736 | 0.1383 | 4.42 | 18.0 | 0.9065 | 0.3684 |
| 3.10 | 15.8 | 0.9797 | 0.1567 | 4.43 | 18.4 | 0.6945 | 0.4363 |
| 3.13 | 11.8 | 0.9913 | 0.0663 | 4.52 | 20.9 | 0.2089 | 0.7220 |
| 3.18 | 13.1 | 0.9876 | 0.1212 | 4.56 | 17.1 | 0.9152 | 0.3384 |
| 3.21 | 12.4 | 0.9783 | 0.0903 | 4.59 | 15.4 | 0.9283 | 0.2931 |
| 3.25 | 13.7 | 0.9759 | 0.1890 | 4.63 | 17.6 | 0.7535 | 0.5529 |
| 3.28 | 14.5 | 0.9905 | 0.0803 | 4.66 | 17.8 | 0.5753 | 0.7426 |
| 3.29 | 18.8 | 0.0523 | 0.9539 | 4.70 | 19.7 | 0.5964 | 0.7249 |
| 3.31 | 13.9 | 0.8355 | 0.3275 | 4.74 | 19.1 | 0.8828 | 0.4334 |
| 3.36 | 18.4 | 0.8843 | 0.3584 | 4.77 | 17.5 | 0.7263 | 0.5718 |
| 3.40 | 12.8 | 0.9831 | 0.1340 | 4.87 | 18.9 | 0.8851 | 0.4264 |
| 3.43 | 14.0 | 0.9832 | 0.1376 | 4.92 | 15.2 | 0.5966 | 0.7401 |
| 3.45 | 13.3 | 0.9835 | 0.1479 | 4.95 | 18 7 | 0.7585 | 0.5930 |
| 3.48 | 15.6 | 0.9899 | 0.0756 | 5.00 | 13.5 | 0.9757 | 0.1938 |
| 3.54 | 15.6 | 0.9837 | 0.1506 | 5.21 | 19.0 | 0.2834 | 0.7163 |
| 3.56 | 17.3 | 0.8835 | 0.3470 | 5 37 | 13.3 | 0.8166 | 0.2957 |
| 3.59 | 13.4 | 0.9888 | 0.1374 | 5.46 | 14.3 | 0.9283 | 0.2779 |
| 3.62 | 13.7 | 0.9894 | 0.1070 | 5.54 | 17.1 | 0.8166 | 0.4332 |

*Table 5.* Diversity of benthic foraminifera (standardized number of species $S_{50}$) and loadings of the first (PC1) and second (PC2) varimax rotated Q-mode principal components axes in samples from Hole 503. These account for 81% of the variability in sample space.

| Age (Ma) | Diversity | PC1 | PC2 | Age (Ma) | Diversity | PC1 | PC2 |
|---|---|---|---|---|---|---|---|
| 1.66 | 17.0 | 0.2910 | −0.8999 | 3.20 | 17.7 | 0.6869 | −0.6441 |
| 1.81 | 19.9 | 0.1820 | −0.8509 | 3.22 | 16.4 | 0.6330 | −0.6224 |
| 2.01 | 18.4 | 0.0853 | −0.8481 | 3.28 | 21.3 | 0.8087 | −0.3305 |
| 2.05 | 21.3 | 0.1800 | −0.8113 | 3.29 | 14.9 | 0.8319 | −0.2838 |
| 2.07 | 14.4 | 0.8813 | −0.4027 | 3.31 | 15.5 | 0.9607 | −0.1142 |
| 2.09 | 17.4 | 0.4732 | −0.7349 | 3.33 | 12.9 | 0.8627 | −0.3171 |
| 2.14 | 19.0 | 0.7557 | −0.4869 | 3.34 | 15.2 | 0.7028 | −0.6155 |
| 2.25 | 12.8 | 0.7319 | −0.4765 | 3.36 | 19.3 | 0.5686 | −0.7524 |
| 2.27 | 12.6 | 0.9654 | −0.1059 | 3.38 | 16.2 | 0.8740 | −0.3853 |
| 2.29 | 15.3 | 0.6861 | −0.5861 | 3.40 | 13.9 | 0.7784 | −0.4059 |
| 2.33 | 15.2 | 0.8235 | −0.3061 | 3.42 | 16.1 | 0.9732 | −0.0625 |
| 2.38 | 18.0 | 0.7828 | −0.4971 | 3.44 | 14.8 | 0.8131 | −0.3222 |
| 2.42 | 16.9 | 0.7036 | −0.6268 | 3.45 | 18.1 | 0.5933 | −0.2707 |
| 2.46 | 17.7 | 0.7285 | −0.5999 | 3.46 | 15.5 | 0.7775 | −0.5190 |
| 2.57 | 15.7 | 0.4081 | −0.7822 | 3.46 | 19.5 | 0.8286 | −0.3979 |
| 2.59 | 13.3 | 0.9491 | −0.0710 | 3.47 | 22.7 | 0.6000 | −0.6728 |
| 2.61 | 19.5 | 0.8618 | −0.4467 | 3.50 | 16.5 | 0.8031 | −0.4867 |
| 2.63 | 16.3 | 0.8951 | −0.3565 | 3.63 | 13,0 | 0.0437 | −0.7718 |
| 2.69 | 17.8 | 0.8387 | −0.4548 | 3.79 | 14.7 | 0.6477 | −0.5907 |
| 2.78 | 21.3 | 0.9361 | −0.2528 | 3.84 | 20.2 | 0.6200 | −0.5946 |
| 2.79 | 21.6 | 0.8285 | −0.3538 | 3.87 | 18.3 | 0.8142 | −0.2666 |
| 2.84 | 20.4 | 0.8632 | −0.4180 | 3.92 | 19.4 | 0.8253 | −0.4805 |
| 2.87 | 18.9 | 0.9258 | −0.1033 | 3.95 | 19.1 | 0.9750 | −0.0631 |
| 2.89 | 20.0 | 0.6577 | −0.6208 | 4.11 | 16.8 | 0.7173 | −0.4451 |
| 2.99 | 17.8 | 0.8291 | −0.4555 | 4.21 | 23.0 | 0.0935 | −0.6809 |
| 3.03 | 21.0 | 0.7870 | −0.4243 | 4.25 | 17.9 | 0.5490 | −0.7450 |
| 3.07 | 17.0 | 0.7322 | −0.3298 | 4.27 | 23.1 | 0.1678 | −0.7015 |
| 3.11 | 18.0 | 0.9105 | −0.2945 | 4.28 | 21.0 | 0.8641 | −0.2264 |
| 3.13 | 15.2 | 0.9580 | −0.2323 | 4.30 | 20.7 | 0.7982 | −0.4261 |
| 3.14 | 17.0 | 0.7615 | −0.5295 | 4.80 | 15.0 | 0.1101 | −0.8479 |
| 3.17 | 19.7 | 0.5979 | −0.5767 | 4.90 | 18.4 | 0.2669 | −0.7120 |
| 3.19 | 17.5 | 0.6318 | −0.5306 | 5.07 | 20.3 | 0.5943 | −0.7438 |

# Results

## Coarse-fraction analysis

Variations in coarse-fraction percentages in Holes 502A and 503 are shown in Figs. 2 and 3. In Hole 502A the coarse fraction decreases slightly to about 4.0 Ma (Fig. 2). This is followed by an increasing trend throughout the remaining parts of the sequence. In Hole 503 the coarse fraction fluctuates around 10%, except for a few intervals at 3.9, 3.8, 2.8, 2.5, and 2.2 Ma, where the percentage increases to about 75% (Fig. 3).

## Fragmentation patterns

The degree of fragmentation increases from the late Miocene to about 4.5 Ma in Hole 502A and exhibited a pulse-like pattern from 5 to about 3.85 Ma (Fig. 2; Table 1). In the interval between 3.85 and 3.0 Ma, the percentage of fragments decreased, generally to less than 50%, except for a few intervals where values are higher than 50% (Fig. 2). From 2.8 to 2.5 Ma the percentage of fragments is generally greater than 50%, a circumstance that indicates stronger dissolution.

In Hole 503 the planktic foraminiferal fragments are more abundant than in Hole 502A, indicating stronger dissolution in this hole than in Hole 502A. Fragments fluctuate around 50–60% up to about 3.9 Ma, whereafter they vary between 70% and 80% in the interval from 3.9 to 2.7 Ma (Fig. 3), which indicates enhanced dissolution. From 2.7 to 2.4 Ma the percentage decreases to 30%, followed by an increase again to about 60% to the top of the sequence (Fig. 3; Table 2).

## Benthic foraminifer acculation rates (BFAR)

The benthic foraminifer accumulation rates (BFAR) in deep-sea sediments are strongly related to the productivity of the surface ocean (Herguera 1992; Berger & Her-

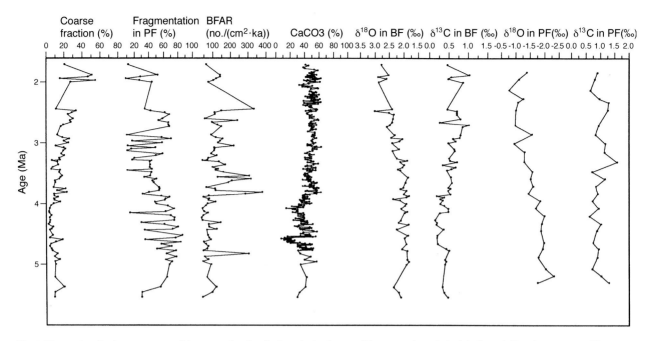

*Fig. 2.* Fluctuations in the percentage of the coarse fraction (>63 μm), the degree of fragmentation of planktic foraminifera (percentage of fragments relative to whole and fragmented tests), BFAR, the accumulation rate of benthic foraminifera per cm² per ka, the percentage of CaCO₃, and δ¹⁸O and δ¹³C of benthic and planktic foraminifera in DSDP Hole 502A. Isotopic measurements are on the epibenthic species *Cibicidoides wuellerstorfi* (>125 m) and the planktic group *Globigerinoides trilobus – G. quadrilobatus – G. sacculifera* (>175 m). The isotopic data of the planktic foraminifera are from Keigwin (1982b). Absolute ages for Hole 502 are from the magnetostratigraphy developed by Kent & Spariosu (1982).

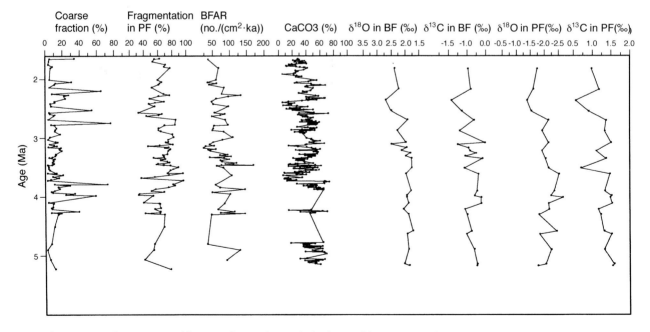

*Fig. 3.* Fluctuations in the percentage of the coarse fraction (>63 μm), the degree of fragmentation of planktic foraminifera (percentage of fragments relative to whole and fragmented tests), BFAR, the accumulation rate of benthic foraminifera per cm² per ka, the percentage of CaCO₃, and δ¹⁸O and δ¹³C of benthic and planktic foraminifera in DSDP Hole 503. Isotopic measurements are on the epibenthic species *Cibicidoides kullenbergi* (>175 m) and the planktic group *Globigerinoides trilobus – G. quadrilobatus – G. sacculifera* (>175 m). The isotopic data of both the benthic and planktic foraminifera are from Keigwin (1982b). Absolute ages for Hole 502 are from the magnetostratigraphy developed by Kent and Spariosu (1982).

guera 1992). As a result, Herguera (1992) and Berger & Herguera (1992) suggested that BFAR may be used to estimate the paleoflux of organic matter to the sea floor ($J_{sf}$) and in turn the surface-water paleoproductivity (*PP*).

In Hole 502A, between the late Miocene and 3.85 Ma, BFAR is in general relatively low (mean = 76), but increased thereafter (mean = 142) (Fig. 2, Table 1). By assuming that the relationships established between BFAR and $J_{sf}$ and *PP* in the modern equatorial ocean hold for the Pliocene ocean and that BFAR is exclusively controlled by surface-water productivity, $J_{sf}$ and *PP* for the pre- and post-3.85 Ma intervals were estimated using the data-modelling technique suggested by Herguera (1992). The pre-3.85 Ma BFAR average may be converted to a $J_{sf}$-value of 110 mg C/(m$^2$·yr), and the post-3.85 Ma value to a $J_{sf}$ of about 180 mg C/(cm$^2$·yr). Similarly, the BFAR for the pre-3.85 Ma period would correspond to a *PP*-value of about 70 gC/(cm$^2$·yr) (in the iterative modelling, the *k*- and *r*-values, which are coefficients in the equations expressing the modern $J_{sf}$, are converted to 17 and 0.6, respectively) and to the period post-3.85 Ma to a *PP*-value equal to 105 gC/(cm$^2$·yr) (*k* converted to 26 and *r* to 0.5). Hence, based on these estimates, the flux of organic matter to the sea floor as reflected in Hole 502A would have been on an average about 60% higher after 3.85 Ma than earlier.

In Hole 503 the BFAR pattern is not as clear as it is in Hole 502A. However, in contrast to Hole 502A, BFAR in Hole 503 shows a higher mean value (86) in the interval from the early Pliocene to about 3.2 Ma and a subsequent decrease to a lower mean value (about 64) (Fig. 3; Table 2).

## Calcium carbonate content

In Hole 502A the CaCO$_3$ content varies between 30% and 55%, showing a gradual increase from about 7.5–3.85 Ma (Fig. 2). From about 3.85 Ma to the top of the sequence the average CaCO$_3$ content is relatively constant (generally 40–50%).

In Hole 503, the general trend of the CaCO$_3$ record consists of a long-wavelength fluctuation throughout the sequence. From about 7.5 to 4.0 Ma the average CaCO$_3$ concentration is approximately 50% (Fig. 3). Between 3.7 and 3.6 Ma it decreases to very low values. Thereafter it increases but fluctuates strongly with intervals of very low CaCO$_3$ content, particularly between 2.5 and 2.4 Ma and at about 1.9 Ma. The average CaCO$_3$ concentration from about 3.6 Ma to the top of the sequence is about 40%.

## Stable isotopes

The signal in the benthic oxygen isotopes exhibits little change before about 3.85 Ma in Hole 502A (Fig. 2; Table 1). Thereafter $^{18}$O becomes gradually enriched throughout the remaining sequence. Between 3.85 Ma and 1.71 Ma, the change in δ$^{18}$O is about 1‰. The δ$^{18}$O of planktic foraminifers shows a similar trend, but here the change begins at about 4.2 Ma. The increase in planktic foraminifer δ$^{18}$O in the interval between 4.2 Ma and the top of the sequence is also about 1‰.

The δ$^{13}$C of benthic foraminifera begins to increase at about 3.85 Ma and continues to the youngest part of the sequence. This change reaches about 0.5‰, whereas no similar trend can be seen in the planktic δ$^{13}$C record.

In Hole 503, the benthic foraminifer $^{18}$O reveals little change before about 3.2 Ma (Fig. 3). Thereafter $^{18}$O becomes gradually enriched throughout the remaining parts of the sequence. A similar trend can be seen in the planktic δ$^{18}$O record, where the δ$^{18}$O values increase by about 0.75‰ in the interval between 3.2 Ma and the top of the sequence.

In contrast to Hole 502A, no discernible trend can be detected in the δ$^{13}$C signal of the benthic foraminifera, although the planktic foraminifer δ$^{13}$C record decreases by about 0.5‰ from approximately 3.0 Ma to the youngest part of the sequence (Fig. 3).

## Diversity of the benthic foraminifer faunas

Fig. 4 shows the relationship between sample size (*N*) and number of species (*S*) found in samples from Hole 502A and Hole 503, respectively. The number of species is generally greater in Hole 502A than in Hole 503, but some portion of the differences could be due to generally larger sample sizes in Hole 502A. The number of species per sample in Hole 502A is 16–44 (mean = 30), whereas in Hole 503 it is 13–33 (mean = 21). As expected, the number of species increases with increasing sample size in both holes (Fig. 4), from between 12 and 25 for a sample size of about 50 specimens to a maximum of 43 for a sample with 448 specimens.

Plots of the Hurlbert diversity index against sample size show that standardization efficiently removes the influence of differences in sample size upon diversity (compare Holes 502A and 503 in Fig. 5). The standardized diversity ranges between 10 and 23 species in Hole 502A (mean diversity is 17) and between 13 and 23 species in Hole 503 (mean diversity is 17), which indicates that there is no major difference in diversity between the holes when sample size is accounted for. Fig. 6 shows the changes in benthic foraminifer diversity ($S_{50}$) throughout the sec-

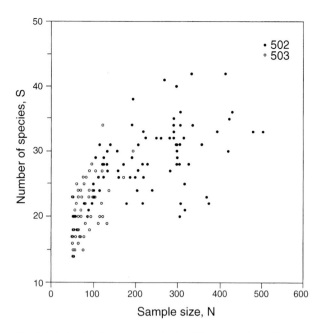

Fig. 4. Relationship between sample size (*N*) and number of benthic foraminifer species (*S*) in Hole 502A (marked with solid circle) and Hole 503 (marked with open circle).

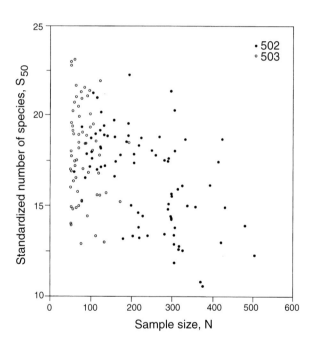

Fig. 5. Relationship between sample size (*N*) and standardized number of benthic foraminifer species ($S_{50}$) in Hole 502A (marked with solid circle) and Hole 503 (marked with open circle).

tions analyzed in Holes 502A and 503 ($S_{50}$ values are listed in Tables 4 and 5). In Hole 502A, the $S_{50}$ value increases from about 15 at 5.5 Ma to about 22–23 at 3.9 Ma and decreases to generally lower values (10–12) in the interval

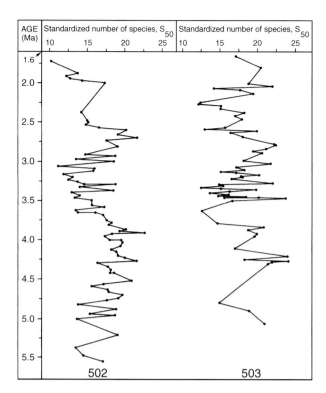

Fig. 6. Variations in diversity of benthic foraminifera (standardized number of species, $S_{50}$) in DSDP Holes 502A and 503.

between 3.8 and 3.0 Ma. Following this interval, the diversity increases to a maximum of 22 at 2.7 Ma and decreases conspicuously to a minimum of 10 at 1.7 Ma. Diversity does not show a similar pattern of variation in Hole 503 (Fig. 6). It decreases from 23 to 13 taxa between 4.2 and 3.6 Ma. This was followed by an increase to 22 taxa between 3.6 and 2.8 Ma and a decrease to 12 at 2.2–2.3 Ma. In the uppermost part of the sequence the diversity increased again to values between 14 and 22.

## Faunal composition

The benthic foraminifer species identified in Holes 502A and 503 and their absolute abundances are presented in Appendix I. Several species are common throughout the interval studied in each hole. In Hole 502A, these include: *Cibicidoides kullenbergi, C. mundulus, C. robertsonianus, C. wuellerstorfi, Epistominella exigua, Gyroidina neosoldanii, Gyroidinoides orbicularis, Laticarinina pauperata, Nuttallides umbonifera, Oridorsalis umbonatus, Pyrgo murrhina* and *Uvigerina peregrina*. In Hole 503, the following species are the most common: *Chilostomella oolina, C. kullenbergi, C. robertsonianus, C. wuellerstorfi, E. exigua, Globocassidulina subglobosa, G. neosoldanii, Melonis pompilioides, N. umbonifera, O. umbonatus, Pullenia bulloides, P. murrhina,* and *Quinqueloculina weaveri*. The dominant species in both holes is *N. umbonifera,* with an absolute abundance ranging between 0 and 266 individuals per sample (mean = 80.3) in Hole 502A and between 0 and 58 individuals per sample (mean = 14.9) in Hole 503.

Agglutinated taxa are relatively rare, generally constituting less than 5% of the fauna. The most common agglutinated species in Hole 502A is *Sigmoilinopsis schlumbergeri,* whereas *Eggerella bradyi* is the most common in Hole 503.

## Faunal differences between the holes

To determine whether there are significant differences in mean absolute abundance of each of the 15 most abundant species from Holes 502A and 503, *t*-test was used (Table 6). The results show differences between the holes for 10 of the species (Table 6). *Cibicidoides kullenbergi, C. robertsonianus, C. wuellerstorfi, G. orbicularis, L. pauperata, N. umbonifera, O. umbonatus,* and *P. murrhina* exhibit greater mean absolute abundance in Hole 502A, whereas two species, *G. subglobosa* and *P. bulloides,* are more abundant in Hole 503.

## Q-mode principal components analysis

*Q*-mode principal components analysis was used to determine (1) benthic foraminifer faunal changes within Holes 502A and 503, and (2) whether the faunal compositions show patterns of change that may be related to the closure of the Isthmus of Panama.

*Faunal changes in Hole 502A.* – The first two varimax-rotated principal components of the Caribbean hole based on fluxes of the 12 most common species (Table 7) account for 90% of the total variability in sample space (the first axis explains 75% and the second represents 15%). The first *Q*-mode score axis divides *N. umbonifera* from the other species, which are neutral along this first axis (Fig. 7; Table 7). The second axis separates *C. wuellerstorfi* and *Oridorsalis umbonatus* from a cluster of the other species.

Fig. 8 illustrates the changes in loadings of the first and second principal component axes plotted against time (cf. Table 4). The first axis shows the variation in flux of *N. umbonifera,* while the second axis shows those of *C. wuellerstorfi* and *O. umbonatus.* The variation along the first principal component axis indicates that *N. umbonifera* is very abundant from 5.0 to about 4.3 Ma, except for a few

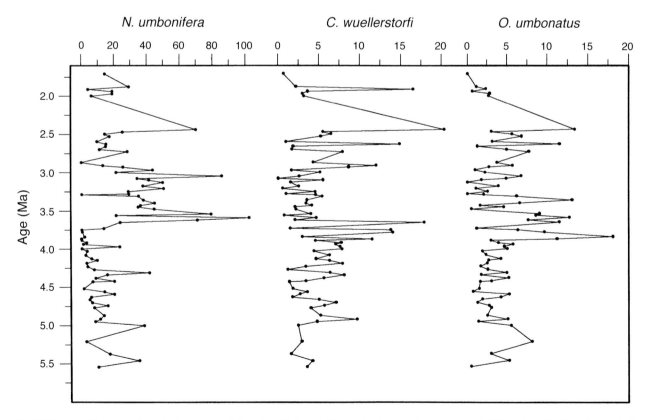

*Fig. 9.* Changes with time in absolute abundances of those benthic foraminifer species that were interpreted as significant from the *Q*-mode principal component analysis (Fig. 7) in Hole 502A.

intervals. This species is also abundant between 3.7 and 3.0 Ma and from about 2.7 to the top of the sequence, with the exception of samples from 3.3, 2.9 and 1.9 Ma.

The second principal component axis shows that *C. wuellerstorfi* and *O. umbonatus* are abundant in the interval between about 4.2 and 3.7 Ma and from 2.9 to 2.8 Ma.

The changes in flux over time are associated with the first two principal components for these three species shown in Fig. 9.

*Faunal changes in Hole 503.* – The first two varimax-rotated principal components of the eastern equatorial

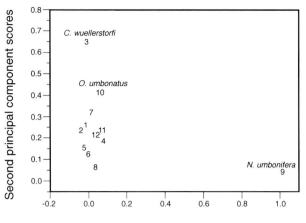

First principal component scores

*Fig. 7.* Plot illustrating the way in which the 12 species (Table 4) included in the Q-mode principal component analysis of Hole 502A are distributed along the first two principal component axes. 1. *Cibicidoides kullenbergi*, 2. *Cibicidoides robertsonianus*, 4. *Epistominella exigua*, 5. *Gyroidina neosoldanii*, 6. *Gyroidinoides orbicularis*, 7. *Laticarinina pauperata*, 8. *Melonis pompilioides*, 11. *Pyrgo murrhina*, 12. *Sigmolinopsis schlumbergeri*,

*Table 6.* Averages ($\bar{x}$) and ranges of absolute abundances of the 12 most common species used in Q-mode principal components analysis of Hole 502A (marked with ☆) and of those 13 species used in Q-mode principal components analysis of Hole 503B (marked with ★), and *t*-tests of differences in the mean absolute abundances between the holes. In those cases where the variances of the absolute abundance were heterogeneous (marked by a significant *F*-value), the alternative *t*-test described by Sokal & Rohlf (1969. p. 146) was employed.

| Species | Hole 502A $\bar{x}$ | Hole 502A Range | Hole 503B $\bar{x}$ | Hole 503B Range | F | t |
|---|---|---|---|---|---|---|
| ☆ ★ *Cibicidoides kullenbergi* | 6.05 | 0–27 | 2.2 | 0–13 | 6.2[c] | 5.2[c] |
| ☆ *Cibicidoides robertsonianus* | 6.5 | 0–27 | 1.1 | 0–16 | 4.3[c] | 8.5[c] |
| ☆ ★ *Cibicidoides wuellerstorfi* | 20.5 | 0–155 | 4.7 | 0–21 | 21.8[c] | 6.4[c] |
| ☆ ★ *Epistominella exigua* | 11.2 | 0–115 | 6.9 | 0–39 | 6.9[c] | 1.8[a] |
| ★ *Globocassidulina subglobosa* | 2.6 | 0–33 | 2.8 | 0–13 | 3.3[c] | 0.4 |
| ☆ ★ *Gyroidina neosoldanii* | 6.3 | 0–18 | 3.2 | 0–8 | 3.9[c] | 5.6[c] |
| ☆ ★ *Gyroidinoides orbicularis* | 6.0 | 0–24 | 2.1 | 0–11 | 5.2[c] | 5.8[c] |
| ☆ ★ *Laticarinina pauperata* | 7.3 | 0–29 | 2.5 | 0–8 | 9.2 | 6.6[c] |
| ☆ ★ *Melonis pompilioides* | 2.5 | 0–45 | 3.6 | 0–12 | 3.2 | 1.4 |
| ☆ ★ *Nuttallides umbonifera* | 79.5 | 0–266 | 14.9 | 0–58 | 27.2[c] | 8.4[c] |
| ☆ ★ *Oridorsalis umbonatus* | 15.5 | 0–47 | 8.3 | 0–26 | 5.5[c] | 5.2[c] |
| ★ *Pullenia bulloides* | 1.0 | 0–11 | 1.8 | 0–15 | 2.2[b] | 2.2[a] |
| ☆ *Pyrgo murrhina* | 10.8 | 0–142 | 1.7 | 0–6 | 104.5[c] | 4.8[c] |
| ★ *Quinqueloculina weaveri* | 3.3 | 0–9 | 2.2 | 0–6 | 1.6[a] | 3.1[c] |
| ☆ *Sigmoilinopsis schlumbergeri* | 8.4 | 0–31 | 0 | 0 | 96.8[c] | 39.0[c] |

[a]$p \leq 0.05$; [b]$p \leq 0.01$; [c]$p \leq 0.001$

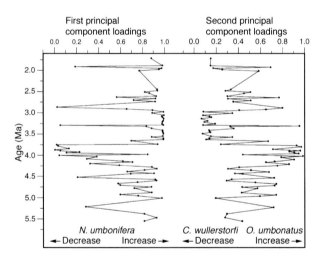

*Fig. 8.* Variations in loadings of the samples along the first and second Q-mode principal component axes in Hole 502A (compare Fig. 7).

*Table 7.* Varimax scores for the 12 species included in the Q-mode principal components analysis of benthic foraminifera from Hole 502A (PC1 represents the first axis and PC2 the second axis.)

| | PC1 | PC 2 |
|---|---|---|
| *Cibicidoides kullenbergi* | −0.0227 | 0.2291 |
| *Cibicidoides robertsonianus* | −0.0124 | 0.2395 |
| *Cibicidoides wuellerstorfi* | −0.0501 | 0.6641 |
| *Epistominella exigua* | 0.0540 | 0.2032 |
| *Gyroidina neosoldanii* | 0.0034 | 0.1692 |
| *Gyroidinoides orbicularis* | 0.0161 | 0.1391 |
| *Laticarinina pauperata* | −0.0199 | 0.2967 |
| *Melonis pompilioides* | 0.0190 | 0.0750 |
| *Nuttallides umbonifera* | 0.9960 | 0.0212 |
| *Oridorsalis umbonatus* | 0.0085 | 0.4238 |
| *Pyrgo murrhina* | 0.0232 | 0.2044 |
| *Sigmoilopsis schlumbergeri* | 0.0128 | 0.2108 |

Pacific Ocean are based on fluxes of the 13 most common species (Table 8), which account for 81% of the total variability in sample space (the first axis explains 72% and the second represents 9%). The first and second Q-mode score axis distinguish *N. umbonifera* and *O. umbonatus* from the other species, which are neutral along this first axis (Fig. 10; Table 8).

The variations in loadings of the first and second principal component axes are plotted against time in Fig. 11 (cf. Table 5). These patterns show the variations in flux of *N. umbonifera* (first axis) and *O.*

*Table 8.* Varimax scores for the 13 species included in the Q-mode principal components analysis of benthic foraminifera from Hole 503 (PC1 represents the first axis and PC2 the second axis.)

| | PC1 | PC2 |
|---|---|---|
| *Cibicidoides kullenbergi* | 0.0263 | −0.1649 |
| *Cibicidoides wuellerstorfi* | 0.1161 | −0.2408 |
| *Epistominella exigua* | 0.2230 | −0.2213 |
| *Globocassidulina subglobosa* | −0.0848 | −0.3652 |
| *Gyroidina neosoldanii* | −0.0202 | −0.3473 |
| *Gyroidinoides orbicularis* | 0.0395 | −0.1342 |
| *Laticarinina pauperata* | 0.0245 | −0.2046 |
| *Melonis pompilioides* | 0.0486 | −0.2510 |
| *Nuttallides umbonifera* | 0.9565 | 0.1278 |
| *Oridorsalis umbonatus* | 0.0800 | −0.6346 |
| *Pullenia bulloides* | 0.0113 | −0.1377 |
| *Pyrgo murrhina* | −0.0152 | −0.1822 |
| *Quinqueloculina weaveri* | 0.0489 | −0.1336 |

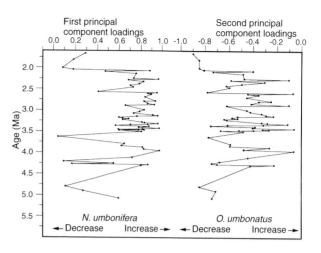

*Fig. 11.* Variations in loadings of the samples along the first and second Q-mode principal component axes in Hole 503 (compare Fig. 10).

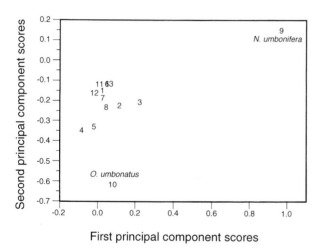

*Fig. 10.* Plot illustrating the way in which the 13 species (Table 5) included in the Q-mode principal component analysis of Hole 502A are distributed along the first two principal component axes. 1. *Cibicidoides kullenbergi*, 2. *Cibicidoides wuellerstorfi*, 3. *Epistominella exigua*, 4. *Globocassidulina subglobosa*, 5. *Gyroidina neosoldanii*, 6. *Gyroidinoides orbicularis*, 7. *Laticarinina pauperata*, 8. *Melonis pompilioides*, 11. *Pullenia bulloides*, 12. *Pyrgo murrhina*, 13. *Quinqueloculina weaveri*.

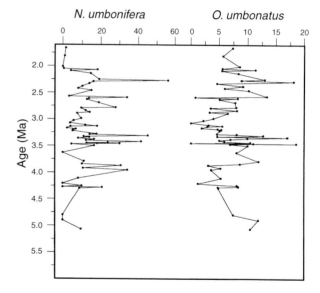

*Fig. 12.* Changes with time in absolute abundances of those benthic foraminifer species that were interpreted as significant from the Q-mode principal component analysis (Fig. 10) in Hole 503.

*umbonatus* (second axis). *Nuttallides umbonifera* is abundant in the interval between 4.3 and 3.8, with the exception of the samples at 4.21 and 4.27 Ma. This species is also abundant between 3.6 and 2.7 Ma and from about 2.5 to 2.1 Ma.

The variation along the second principal component axis indicates that *O. umbonatus* is abundant in the inter-val between about 5.0 and 4.8 Ma and at about 4.2 Ma. This species is also abundant between 3.8 and 3.3 Ma, and at 2.5 Ma and from about 2.0 to the top of the sequence, with the exception of a few samples in between (Fig. 11).

The changes in the flux over time are associated with the first two principal components for *N. umbonifera* and *O. umbonatus* (Fig. 12).

# Paleoceanographic interpretations

The planktic foraminifera became gradually enriched in [18]O beginning at about 4.2 Ma in the Caribbean Hole 502A, whereas no similar change was noted in the benthic record until 0.35 Ma later, at about 3.85 Ma (Fig. 2). Considering that there is no faunal evidence for any major cooling of Caribbean surface waters as early as 4 Ma, this may reflect increasing surface-water salinity in the Caribbean Sea as a result of restricted surface-water communication between the Atlantic and Pacific oceans, caused by the emergence of the Panamanian Isthmus (Keigwin 1982b–c).

At about 3.85 Ma, most of the physico-chemical variables changed, as indicated by changes in the coarse fraction, dissolution (fragmentation of planktic foraminifera), BFAR, $CaCO_3$, $\delta^{18}O$ and $\delta^{13}C$ of benthic foraminifera, and was probably caused by an exchange of the bottom-water inflow into the Caribbean from the North Atlantic. The results of the t-tests confirm that a difference exists between the pre- and post-3.85 Ma periods in all variables except $\delta^{13}C$ of planktic foraminifera (Table 9). The improved $CaCO_3$ preservation, as well as greater proportions of planktic foraminifer-bearing coarse fraction and lesser fragmentation in planktic foraminifera at 3.85 Ma (Fig. 2), point to reduced dissolution and improved calcite preservation. This is probably partly caused by a better ventilation of the bottom waters due to the bottom-water exchange, and partly by the uplift of Site 502 above the lysocline at this time (Gardner 1982). The increased BFAR at about 3.85 Ma may be due to increased productivity in the surface waters in the Caribbean Sea but could also be a result of improved calcite preservation or better ventilation of the bottom waters.

Q-mode principal component analysis based on flux rates of benthic foraminifera distinguished two faunal groups in Hole 502A, one of which is composed of *N. umbonifera* and the other of *C. wuellerstorfi* and *O. umbonatus* (Fig. 7). These species showed the overall highest accumulation rates. From 4.2 to 3.7 Ma the flux of *N. umbonifera* decreased while that of *C. wuellerstorfi* and *O. umbonatus* increased.

Earlier work correlates some benthic foraminifers with specific water masses in the oceans (e.g., Streeter 1973; Schnitker 1974; Boltovskoy 1976; Lohmann 1978, for the Atlantic Ocean; Belanger & Streeter 1980, for the Norwegian–Greenland Sea; Corliss 1979, for the Indian Ocean; Boltovskoy 1976; Ingle *et al.* 1980, for the southeast Pacific Ocean). *Cibicidoides wuellerstorfi* and *O. umbonatus* have been regarded as indicator species of NADW (Lohmann 1978, 1981; Weston & Murray 1984; Hermelin 1986, 1989; Murray 1991; among others). *Nuttallides umbonifera* has been associated with AABW (Streeter 1973; Schnitker 1974; Lohmann 1978, 1981; Corliss 1978, 1979, 1983; Hodell *et al.* 1985; Hermelin 1986, 1989; Murray 1991; among others). However, the relationships between benthic foraminifer species and properties of the bottom-water masses is still unclear (Thomas 1985; B.H. Corliss, personal communication, 1994). Since AABW does not normally flow at depths less than 4,000 m in the world oceans today, an inflow of AABW into the Caribbean Sea during the late Neogene must therefore be regarded as highly unlikely.

*Nuttallides umbonifera* has been found to be related to bottom waters that are highly corrosive to $CaCO_3$, and thus with higher concentrations of dissolved $CO_2$ (Bremer & Lohmann 1982; Mackensen *et al.* 1990). Moreover, Miller (1983) and Tjalsma & Lohmann (1983) reported that *N. umbonifera* is abundant in old, sluggish, oxygen-poor bottom waters, whereas Woodruff & Doug-

*Table 9.* Tests (*t*-tests) of the differences in mean values of the various physico-chemical variables between the intervals 5.54–3.85 Ma and 3.85–1.71 Ma in Hole 502A. $N$ is the sample size, $\bar{x}$ the mean value, and $s$ the standard deviation. The $F$-value results from a standard variance ratio test of the homogeneity of variances. In cases of homogeneous variances the ordinary Student's *t*-test (marked by 'o') was employed to test the significance of differences in means, whereas the alternative $t_s$-value (marked by 's') was computed when variances were heterogeneous (Sokal & Rohlf 1969, pp. 374–375). The $CaCO_3$ data are from Gardner (1982), and the isotope data of planktic foraminifera are from Keigwin (1982b).

| | | 5.54–3.85 Ma | | | 3.85–1.71 Ma | | | |
| --- | --- | --- | --- | --- | --- | --- | --- | --- |
| | $N$ | $\bar{x}$ | $s$ | $F$ | $\bar{x}$ | $s$ | $F$ | $t$ |
| Coarse fraction (%) | 35 | 7.70 | 4.30 | 45 | 19.30 | 10.80 | 6.23[c] | 6.56(s)[c] |
| Fragmentation in planktic foraminifera (%) | 35 | 60.30 | 17.50 | 45 | 41.00 | 18.30 | 1.10 | 480(o)[c] |
| BFAR (No./(cm²·ka)) | 35 | 76.00 | 46.90 | 45 | 142.40 | 77.00 | 270[b] | 4.76(s)[c] |
| $CaCO_3$ (%) | 114 | 34.80 | 10.60 | 228 | 49.30 | 6.00 | 310[c] | 13.63(s)[c] |
| $\delta^{18}C$ in benthic foraminifera (‰) | 24 | 2.03 | 0.17 | 33 | 2.32 | 0.30 | 3.31[b] | 4.62(s)[c] |
| $\delta^{13}C$ in benthic foraminifera (‰) | 24 | 0.32 | 0.12 | 33 | 0.61 | 0.17 | 2.28[b] | 7.65(s)[c] |
| $\delta^{18}C$ in planktic foraminifera (‰) | 16 | −1.92 | 0.19 | 15 | −1.25 | 0.26 | 1.82 | 8.22(o)[c] |
| $\delta^{13}C$ in planktic foraminifera (‰) | 16 | 0.98 | 0.23 | 15 | 1.00 | 0.25 | 1.16 | 0.19(o) |

[a]$p \leq 0.05$; [b]$p \leq 0.01$; [c]$p \leq 0.001$

las (1981) concluded that this species is indicative of young, oxygenated bottom waters in the western Pacific. According to Gooday (1993), it may also be instructive to compare the distribution of *N. umbonifera* with that of *E. exigua*, which is controlled largely by the presence of organic material at the seafloor. However, *N. umbonifera* is abundant south of 40°N in the North Atlantic (Weston & Murray 1984), in areas which probably receive a minimal phytodetrital input (Gooday 1993), and, in addition, *E. exigua* shows relatively low abundance in the interval between 3.7 and 3.0 Ma, where *N. umbonifera* is abundant in the Caribbean Sea. According to Lukashina (1988), *N. umbonifera* can occur as far north as 55°–57°N. The abundances are low (<5%) in the Northeast Atlantic, where *E. exigua* is abundant, but *N. umbonifera* reach 31–55% in the Northwest Atlantic, where *E. exigua* is less common. *Nuttallides umbonifera* and *E. exigua* co-occur in the Weddell Sea and South Atlantic, with *E. exigua* dominating above the lysocline and *N. umbonifera* below the lysocline and above the CCD (Mackensen *et al.* 1990, 1993). Gooday (1993) found that at Discovery Stn 12174 in the Northeast Atlantic, several specimens of *N. umbonifera* had dark green protoplasm packed with algal cells (visible after the test had been dissolved away in glycerol), indicating a diet similar to that of *E. exigua*. Although they have a similar diet, *N. umbonifera* is a non-opportunist, able to survive, when necessary, on a lower food supply than *E. exigua* (Gooday 1993). *Nuttallides umbonifera* has a relatively large, thick-walled test, which suggests a slow rate of growth compared with the small, thin-walled tests of *E. exigua*. As a result, *N. umbonifera* could be outcompeted, or numerically swamped, by a fast-growing opportunist such as *E. exigua* in areas of phytodetritus deposition. Therefore, it could be that *N. umbonifera* has more chance to flourish where the scarcity of food (Loubere 1991) or the corrosive nature of the bottom water (Bremer & Lohmann 1982; Mackensen *et al.* 1990) make conditions less favourable for *E. exigua* and similar species.

However, undersaturation of bottom water is not the factor at the site of the Caribbean Hole 502A, because there is no evidence of an increase in calcite dissolution at 3.7 Ma. The increasing BFAR values after about 3.85 Ma suggest that the supply of organic matter to the sea-floor increased and, therefore, the availability of food was relativily high during the period when *N. umbonifera* was abundant. However, according to B. Corliss (personal communication, 1994), *N. umbonifera, C. wuellerstorfi* and *O. umbonatus* are all dwelling in deep-sea environments with relatively low food supplies. In addition, there is no clear correlation between changes in BFAR and faunal composition in the site range studied. Therefore, it is most likely that the increased abundance of *N. umbonifera* was caused by a change in other environmental conditions in the Colombia Basin. These may be related to the changes in the organic-carbon content and/or to physico-

chemical parameters (e.g., salinity, temperature, oxygen content), but it seems most unlikely that it was caused by the corrosive nature of the bottom water.

A rapid change in the benthic foraminifer fauna occurred at about 3.7 Ma, when *N. umbonifera* became the most abundant benthic species and *C. wuellerstorfi* and *O. umbonatus* decreased in abundance (Figs. 8–9). This event does not coincide with the change in the physico-chemical variables at 3.85 Ma (Fig. 2) but may be a response to a change in the bottom-water circulation between the North Atlantic and the Caribbean Sea. The bottom-water exchange in the Caribbean Sea is probably related to an increased northward transport of warm, high-salinity waters to high latitudes via the Gulf Stream, caused by the progressive emergence of the Panamanian Isthmus (Hodell *et al.* 1985). As the Gulf Stream water is considered to be an important component in the North Atlantic surface circulation, this warm, high-salinity water may have stimulated production of LNADW (Worthington 1970) in the North Atlantic, beginning at about 3.85 Ma. This is because LNADW formation is mainly controlled by salinity differences (deMenocal *et al.* 1992; Raymo *et al.* 1992; Lehman & Keigwin 1992). The LNADW would expand vertically to greater depth in the Atlantic Ocean, thereby preventing it from flowing over the 1,650-m Windward Passage sill into the Colombia Basin. This would in turn favor reduced inflow of UNADW into the Caribbean Sea (deMenocal *et al.* 1992). However, there is no simple compensatory relationship between LNADW and UNADW circulation, which does not exclude the possibility that there might be some production of UNADW with increased salinity as well (deMenocal *et al.* 1992). The inflowing bottom-water entering the western Caribbean Sea over the the Windward Passage sill today consists almost exclusively (85%) of UNADW (deMenocal *et al.* 1992). Thus, the physico-chemical changes, and probably also the faunal changes, could be a result of increased inflow of UNADW into the Caribbean Sea at 3.85 Ma. The relatively high nutrient content in the deep water of the Colombia Basin during this period may indicate, at least in part, that a mixing process between the base of the high nutrient-rich Antarctic Intermediate Water (AAIW) and UNADW (Haddad *et al.* 1994) or local outflow (e.g., from the Rio Magdalena) could have been present as well.

The physico-chemical variables, the coarse fraction, dissolution, BFAR, $CaCO_3$, $\delta^{18}O$ and $\delta^{13}C$ of planktic and benthic foraminifera in the eastern equatorial Pacific Hole 503 show a similar trend to Hole 502A until about 3.8 Ma (Figs. 2–3; Table 10). After 3.8 Ma, the two areas respond differerently: Hole 503 continued with a typical equatorial Pacific character as seen in the $CaCO_3$ record, while the record of Hole 502A was affected by local tectonic events as well as more distant sources, such as the bottom-water exchange between the Caribbean Sea and

*Table 10.* Tests (*t*-tests) of the differences in mean values of the various physico-chemical variables between the intervals 5.23–3.85 Ma and 3.85–1.65 Ma in Hole 503. *N* is the sample size, $\bar{x}$ the mean value, and *s* the standard deviation. The *F*-value results from a standard variance ratio test of the homogeneity of variances. In cases of homogeneous variances the ordinary Student's *t*-test (marked by '*o*') was employed to test the significance of differences in means, whereas the alternative $t_s$-value (marked by '*s*') was computed when variances were heterogeneous (Sokal & Rohlf 1969, pp. 374–375). The $CaCO_3$ data are from Gardner (1982), and the isotope data of benthic and planktic foraminifera are from Keigwin (1982b).

| | 5.23–3.85 Ma | | | 3.85–1.65 Ma | | | | |
| --- | --- | --- | --- | --- | --- | --- | --- | --- |
| | *N* | $\bar{x}$ | *s* | *N* | $\bar{x}$ | *s* | *F* | *t* |
| Coarse fraction (%) | 25 | 14.80 | 13.00 | 71 | 14.80 | 13.00 | 1.279 | 0.467 |
| Fragmentation in planktic foraminifera (%) | 22 | 57.50 | 13.90 | 66 | 67.20 | 12.70 | 1.206 | 2.882 |
| BFAR (No./(cm²·ka)) | 13 | 87.20 | 36.70 | 51 | 70.90 | 28.50 | 1.663 | 1.496 |
| $CaCO_3$ (%) | 42 | 49.50 | 13.80 | 283 | 37.90 | 15.10 | 1.193 | 5.012 |
| $\delta^{18}O$ in benthic foraminifera (‰) | 22 | 1.90 | 0.10 | 10 | 2.30 | 0.27 | 6.354 | 4.147 |
| $\delta^{13}C$ in benthic foraminifera (‰) | 22 | −0.26 | 0.28 | 10 | −0.52 | 0.26 | 1.224 | 2.624 |
| $\delta^{18}O$ in planktic foraminifera (‰) | 22 | −0.25 | 0.28 | 10 | −0.519 | 0.255 | 1.224 | 2.624 |
| $\delta^{13}C$ in planktic foraminifera (‰) | 22 | 1.42 | 0.30 | 10 | 1.11 | 0.30 | 1.023 | 2.545 |

[a]$p \leq 0.05$; [b]$p \leq 0.01$; [c]$p \leq 0.001$

the Atlantic, Rio Magdalena, etc., all working somewhat independently of each other (Gardner 1982). The amplitudes of the fluctuations in the coarse fraction, dissolution and $CaCO_3$ records are, in general, much larger in Hole 503 than in Hole 502A. According to Gardner (1982), the carbonate record in Hole 503 indicates that the sediments were never below the Calcite Compensation Depth (CCD) but were affected instead by fluctuations in the depth of the lysocline. Therefore, the differences in the records of coarse fraction, preservation of planktic foraminifera, and $CaCO_3$ content could be attributed to dissolution signals caused by the location of the lysocline in Hole 503 (Gardner 1982). It appears, therefore, that small movements of the lysocline may translate into dissolution events. The BFAR record shows a similar trend to the $CaCO_3$ content at Hole 503, which may indicate that this parameter can also be affected by changes of the lysocline, as well as it could be a result of changes in the productivity.

It has been suggested that the global climate was generally warmer and that the Antarctic glacier system did not fluctuate on a large scale during the Pliocene prior to 3.3–3.2 Ma (Crowley 1991; Hodell & Venz 1992). In the Northern Hemisphere, glaciers and sea ice were greatly reduced or absent during this time, and ice-volume fluctuations were largely restricted to Antarctica. From about 3.3–3.2 Ma the climate changed from generally warm conditions to colder, high-amplitude climatic variations, which led up to the onset of the major Northern Hemisphere glaciation at 2.47 Ma (e.g., Shackleton *et al.* 1984; Ruddiman *et al.* 1987). These climatic and large-scale surface and deep circulation changes of the North Atlantic were probably caused by small fluctuations in the North Hemisphere ice-sheet volume or other factors such as closing of the Panamanian Isthmus or opening of the Bering Strait (e.g., Ruddimen *et al.* 1987). Stable-isotope results for benthic and planktic foraminifers from sediment between 2.0 and 4.0 Ma at DSDP Site 606 (on the

west flank of the Mid-Atlantic Ridge, present water depth 3,007 m) indicate $^{18}O$ enrichment at 2.4 and 2.6 Ma, reflecting Northern Hemisphere glacial advances similar to these seen elsewhere in the North Atlantic Ocean (DSDP Hole 552A, water depth 2,301 m) (Keigwin 1986). However, Site 606 also differ significantly from those at Site 552, where oxygen isotopic evidence indicates some combination of increased ice volume and deep-water cooling of NADW between 3.1 and 3.4 Ma (Keigwin 1986). Keigwin (1986) suggested that minor glaciation occurred during the earliest pulse (3.1 Ma), but that this was mostly a deep-sea cooling event and, therefore, not reflected in the shallower located Site 552. This permanent $^{18}O$ enrichment at about 3.1 Ma is comparable to earlier estimates in Pacific benthic foraminifers (Prell 1984). Although minor glaciers had reached the North Atlantic and produced some ice rafting prior 2.57 Ma, the lack of significant ice-rafted debris in the high latitude North Atlantic also argues against large continental ice sheets on the Northern Hemiphere before that time (Jansen & Sjøholm 1991).

Planktic and benthic foraminifer oxygen isotopic records from Hole 503 show little change before 3.2 Ma (Fig. 3). This is contrary to what is observed in Hole 502A. Thereafter $^{18}O$ in Hole 503 became gradually enriched and remained so, with a few exceptions, throughout the sequence studied. In addition, the lack of a similar oxygen isotopic signal in the Caribbean Hole 502A at about 3.2 Ma supports the argument that the enriched $^{18}O$ in Hole 503 at 3.2 Ma was a result of climatic and oceanographic changes that mainly affected deep-sea areas area (e.g., Prell 1984; Keigwin 1986; Ruddiman *et al.* 1986; Hodell & Venz 1992).

The Q-mode principal component analysis shows that *N. umbonifera* was the most abundant benthic species in the interval between 4.8 and 2.1 Ma, with exceptions at 4.3, 4.1, 3.7, and 2.6 Ma (Fig. 11). *Nuttallides umbonifera* was also abundant in the interval between 4.8 and 4.1 Ma at other sites in the Pacific Ocean (e.g., Woodruff 1985; Hermelin 1989). Although the distribution pattern of *N. umbonifera* shows some similarities between Hole 503 and the Caribbean Hole 502A the physico-chemical parameters do not support any bottom-water exchange across Panama after 4.0 Ma (Figs. 2–3). According to Thomas (1992), rapid benthic foraminifer faunal changes in the deep sea could be a result of changes in the source area of the deep water masses as well as changes in the character of the waters in the source areas. However, changes in productivity, which in turn might also have been influenced by changing oceanic circulation patterns, may complicate the signal of the faunal changes (Thomas 1992).

The predominant *N. umbonifera* assemblage in the interval between about 4.8 and 2.1 Ma may be a result of fluctuations over time in the depth of the boundaries between water masses at the site of Hole 503 due to changes in volumes of the different water masses (e.g., Oberhänsli *et al.* 1991; Thomas 1992). The strongly dissolved faunas may indicate the presence of a corrosive AABW during most of the Pliocene (McDougall 1985); however, there is no significant correlation between dissolution and the variation in the abundance of *N. umbonifera*. Furthermore, the variation in the $CaCO_3$ content showed no correlation to the variation in the abundance of *N. umbonifera*. A second explanation for the predominance of *N. umbonifera* assemblages could be changes in the productivity, but this seems less likely, as the BFAR value does not show any significant correlation with the variation of *N. umbonifera*. Although there is no directly correlation between the abundance of *N. umbonifera* and the BFAR-value, the location of Hole 503 is a typical Pacific deep-sea site with in general low food supply and

periods associated with carbonate-undersaturated bottom waters. Therefore, it cannot be excluded that the high abundance of *N. umbonifera* could be a result of low food supply and/or corrosive nature of the bottom water. A third explanation could be a transitory shift in environmental preference of *N. umbonifera*. However, as long as the knowledge of the relationship between water masses and their benthic foraminifer assemblages is still in its infancy (Boltovskoy *et al.* 1992), details of ecological controls on deep-sea benthic foraminifera populations are poorly understood. Thus, transitory shifts in environmental preferences of specific species can not be evaluated.

The diversity of the benthic foraminifer faunas decreased as a consequense of increased abundance of *N. umbonifera* at 3.7 Ma in Hole 502 (Fig. 6). Decreased diversity of benthic foraminifera may have been a result of increased input of phytodetritus (Gooday & Lambshead 1989; Lambshead & Gooday 1990). This is because the limited number of species colonizing the detritus and the resulting increase in the abundance of these species causes an overall decline in diversity. It is uncertain whether the high abundance of *N. umbonifera* may be a result of increased food supply from 3.7 Ma in Hole 502A, although BFAR is increasing during the same period.

In Hole 503, the diversity of the benthic foraminifer fauna shows the opposite pattern in the same interval in Hole 502 (Fig. 6), although *N. umbonifera* is fairly frequent between 3.7 and 3.0 Ma. Also the BFAR in Hole 503 shows lower values after 3.8 Ma. Therefore, in Hole 503 the diversity of the benthic foraminifera seems to be primarily controlled by dissolution and/or low deposition of phytodetritus.

# Taxonomic description of selected species

The foraminifer classification used in this study conforms chiefly to that of Loeblich & Tappan (1964, 1988), but taxonomic studies by Barker (1960), McCulloch (1981), Kohl (1985), Van Morkhoven et al. (1986), and Hermelin (1989) have also been considered.

The synonymies include references to original descriptions, name changes, and significant discussions. This is followed by a short description of the species, remarks on their taxonomic position, and information about their ecological preferences (e.g., with regard to water depth, salinity, substrate, and productivity), if available. Their distribution in Holes 502A and 503 are as well presented. Most of the species included in this section are illustrated by scanning electron micrographs, which are presented in Figs. 13–27. The absolute abundances of the 147 species are presented in the Appendix.

## Phylum Protozoa Goldfuss, 1818

## Subphylum Sarcodina Schmarda, 1871

## Class Rhizopodea von Siebold, 1845

## Order Foraminiferida Eichwald, 1830

## Suborder Textulariina Delage & Hérouard, 1896

## Subfamily Recurvoidinae Alekseychik-Mitskevich, 1973

## Genus *Recurvoides* Earland, 1934

### *Recurvoides scitulus* (Brady, 1881)
Fig. 13A–B

*Synonymy.* – □1881 *Haplophragmium scitulum* n.sp. – Brady, Pl. 1.8–1.9. □1990 *Recurvoides scitulus* (Brady) – Hermelin & Shimmield, Pl. 1:8–9.

*Description.* – Test free, streptospiral, with the later portion coiled in a different plane than the early portion; periphery broadly rounded. Chambers 5–6 in the final whorl, the majority with 6 chambers. Sutures distinct, slightly depressed, nearly straight. Wall coarsely agglutinated. Aperture interio-areal, ovate to somewhat elongate.

*Distribution at Sites 502A and 503B.* – *Recurvoides scitulus* is absent at Site 502A, while it is rare with few occurrences at Site 503B.

## Subfamily Textulariidae Ehrenberg, 1838

## Genus *Siphotextularia* Finlay, 1939

### *Siphotextularia catenata* (Cushman, 1911)
Fig. 13C–D

*Synonymy.* – □1911 *Textularia catenata* n.sp. – Cushman, p. 23, Text-figs. 39–40. □1951 *Siphotextularia rolshauseni* n.sp. – Phleger & Parker, p. 4, Pl. 1:23–24. □1953 *Siphotextularia rolshauseni* Phleger & Parker, 1951 – Phleger *et al.*, p. 26, Pl. 5:7. □1971 *Siphotextularia rolshauseni* Phleger & Parker, 1951 – Schnitker, p. 210, Pl. 1:15a–b. □1979 *Siphotextularia catenata* (Cushman) – Corliss, p. 5, Pl. 1:1–2. □1981 *Siphotextularia rolshauseni* Phleger & Parker, 1951 – Cole, p. 36, Pl. 5:7. □1984b *Siphotextularia catenata* (Cushman) – Boersma, Pl. 1:5. □1984 *Siphotextularia catenata* (Cushman) – Murray, Pl. 3:8–9. □1985 *Siphotextularia rolshauseni* Phleger & Parker, 1951 – Hermelin & Scott, p. 217, Pl. 1:6a–7b. □1986 *Siphotextularia rolshauseni* Phleger & Parker, 1951 – Kurihara & Kennett, Pl. 1:3. □1989 *Siphotextularia catenata* (Cushman) – Hermelin, pp. 30–31. □1990 *Siphotextularia catenata* (Cushman) – Ujiié, p. 12, Pl. 1:4a–b; 5a–b.

*Description.* – Test free, elongate, biserial; lobate periphery. Chambers globular, increasing in size rapidly. Sutures depressed. Wall coarsely agglutinated, finely perforate. Aperture subterminal, rounded, in the lower base of the ultimate chamber.

*Remarks.* – *Siphotextularia catenata* was described by Phleger & Parker (1951) as an infraspecific variation of *Siphotextularia rolshauseni* in materials from the Gulf of Mexico. In the North Atlantic, Phleger *et al.* (1953) found that specimens were larger than in the Gulf of Mexico and slightly more compressed. They stated that this form (from the North Atlantic) was identical to Cushman's holotype of *Textularia catenata* from the Atlantic, but it did not resemble Cushman's holotype of *T. catenata* from the western North Pacific, which was a much larger form. No other differences between *S. rolshauseni* and *T. catenata* were noted. Thus, this species differs from Cushman's holotype of *T. catenata* only in the size. According to Corliss (1979), variation in size is not a valid specific taxonomic character in benthic foraminifera. Therefore, *S. rolshauseni* should be regarded as a junior synonym of *S. catenata*. This was also confirmed by Thomas (1985), among others.

*Fig. 13.* Scale bar 100 μm. □A, B. *Recurvoides scitulus* (Brady); A, Side view, Hole 503B, 68.95 m; B, Side view, Hole 503B, 112.72. □C, D. *Siphotextularia catenata* (Cushman); C, Side view, Hole 502A, 47.76 m; D, Side view, Hole 502A, 47.76 m. □E, F. *Eggerella bradyi* (Cushman); E, Apertural view, Hole 502A, 49.67 m; F, Side view, Hole 502A, 49.67 m. □G. *Karreriella bradyi* (Cushman), Side view, Hole 502A, 116.52 m. □H. *Martinottiella communis* (d'Orbigny), Side view, Hole 502A, 80.46 m. □I. *Nummoloculina irregularis* (d'Orbigny), Side view, Hole 503B, 69.99 m. □J. *Quinqueloculina weaveri* Rau, Side view, Hole 502A, 78.45 m. □K, L. *Quinqueloculina venusta* Karrer, K Side-edge view, Hole 502A, 46.87 m; L, Side view, Hole 502A, 46.87 m.

*Ecology.* – In the Gulf of Mexico, *S. rolshauseni* (*S. catenata*) has a bathymetric range from the lower middle bathyal zone down into abyssal water depths (Pflum & Frerichs 1976). Corliss (1979) reported this species from 2,500 to 4,600 m in the southeast Indian Ocean.

*Distribution at Sites 502A and 503B.* – *Siphotextularia catenata* occurs throughout the studied sequence at Site 502A. This species is a common species at Site 502A, with absolute abundances between 0 and 15. At Site 503B it occurs rarely and in scattered samples.

# Family Ataxophragmiidae Schwager, 1877

# Subfamily Globotextulariinae Cushman, 1927

# Genus *Eggerella* Cushman, 1933

## *Eggerella bradyi* (Cushman, 1911)
Fig. 13E–F

*Synonymy.* – □1884 *Verneuilina pygmaea* (Egger) – Brady, p. 385, Pl. 47:4–7. □1911 *Verneuilina bradyi* sp.nov. – Cushman, p. 54, Pl. 55:87. □1937b *Eggerella bradyi* (Cushman) – Cushman, p. 52, Pl. 5:19. □1951 *Eggerella bradyi* (Cushman) – Phleger & Parker, p. 6, Pl. 3:1–2. □1953 *Eggerella bradyi* (Cushman) – Phleger *et al.*, p. 27, Pl. 5:8–9. □1960 *Eggerella bradyi* (Cushman) – Barker, p. 96, Pl. 47:4–7. □1964 *Eggerella bradyi* (Cushman) – Leroy, p. 18, Pl. 1:13–14. □1978 *Eggerella bradyi* (Cushman) – Boltovskoy, Pl. 3:33. □1971 *Eggerella nitens* (Wiesner) – Echols, p. 163. □1971 *Eggerella bradyi bradyi* (Cushman) – Herb, p. 296, Pl. 12:1. □1979 *Eggerella bradyi* (Cushman) – Corliss, p. 5, Pl. 1:3–4. □1980 *Eggerella bradyi* (Cushman) – Keller, Pl. 1:8. □1981 *Eggerella bradyi* (Cushman) – Burke, Pl. 1:6. □1984a *Eggerella bradyi* (Cushman) – Boersma, Pl. 8:1. □1984 *Eggerella bradyi* (Cushman) – Murray, Pl. 1:20–21. □1985 *Eggerella bradyi* (Cushman) – Kohl, p. 32, Pl. 3:3. □1985 *Eggerella bradyi* (Cushman) – Mead, pp. 225–226, Pl. 1:1a–b. □1985 *Eggerella bradyi* (Cushman) – Thomas, p. 676, Pl. 1:4. □1985 *Eggerella bradyi* (Cushman) – Boersma, Pl. 10:1–2. □1986 *Eggerella bradyi* (Cushman) – Belanger & Berggren, p. 331, Pl. 1:1. □1989 *Eggerella bradyi* (Cushman) – Hermelin, p. 32, Pl. 2:1–2.

*Description.* – Test free, conical, early stages trochospiral, later stages triserial; chambers inflated, increasing rapidly in size, so that the last whole one forms more than one-half of the test. Wall smooth, finely agglutinated, finely perforate. Sutures distinct and depressed. Aperture a low thin slit at the base of the ultimate chamber.

*Remarks.* – *Eggerella bradyi* has an agglutinated test, which is composed of calcareous particles cemented together. The surface of the perforated test can vary; most specimens have surface textures that are smooth and polished, but some specimens are smooth with a dull luster. Also the test size can vary as well.

*Ecology.* – *Eggerella bradyi* has been reported from almost all bathyal and abyssal depths of the world oceans. In the Gulf of Mexico, the size of the test increased with the water depth, to a maximum length of about 0.3 mm in the upper bathyal water zone and about 1.0 mm in the middle bathyal and abyssal zone (Pflum & Frerichs 1976). In the southwest Atlantic, *Eggerella bradyi* increases in relative abundance from about 0.6% to 5% between 1,493 and 2,769 m, and up to 5–10% between 2,771 and 3,122 m (Mead 1985). In the southeast Indian Ocean, *Eggerella bradyi* is reported between 2,500 and 4,500 m (Corliss 1979)

*Distribution at Sites 502A and 503B.* – *Eggerella bradyi* occurs throughout the studied sequence at Site 502A, with absolute abundances between 0 and 9. At Site 503B it is rare and exhibits scattered occurrences.

# Genus *Karreriella* Cushman, 1933

## *Karreriella bradyi* (Cushman, 1937)

Fig. 13G

*Synonymy.* – □1937b *Karreriella bradyi* (Cushman) – Cushman, p. 135, Pl. 16:6–11. □1945 *Karreriella bradyi* (Cushman) – Cushman & Todd, p. 8, Pl. 1:20. □1949 *Karreriella bradyi* (Cushman) – Bermúdez, p. 89, Pl. 5:11–16. □1960 *Karreriella bradyi* (Cushman) – Barker, p. 94, Pl. 46:1–4. □1964 *Alvarezina bradyi* (Cushman) – Akers & Dorman, pp. 20–21, Pl. 1:22. □1971 *Karreriella bradyi* (Cushman) – Murray, p. 47, Pl. 16:1–4. □1978 *Karreriella bradyi* (Cushman) – Boltovskoy, Pl. 4:28–29. □1979 *Karreriella bradyi* (Cushman) – Corliss, p. 5, Pl. 1:5–6. □1980 *Karreriella* aff. *bradyi* (Cushman) – Butt, Pl. 7:26. □1980 *Karreriella bradyi* (Cushman) [*sic*] – Keller (part), Pl. 1:10 (not 9). □1981 *Karreriella bradyi* (Cushman) – Cole, p. 44, Pl. 6:5. □1985 *Karreriella bradyi* (Cushman) – Hermelin & Scott, p. 212, Pl. 1:8. □1985 *Karreriella bradyi* (Cushman) – Kohl, p. 32, Pl. 3:4–5. □1985 *Karreriella bradyi* (Cushman) – McDougall, p. 394, Pl. 1:4. □1986 *Karreriella bradyi* (Cushman) – Belanger & Berggren, p. 331, Pl. 1:2a–b. □1986 *Karreriella bradyi* (Cushman) – Boersma, 1986, Pl. 10:5. □1989 *Karreriella bradyi* (Cushman) – Hermelin, p. 33.

*Description.* – Test free, elongate, cylindrical, circular in cross–section, four times as long as broad, early portion trochospiral, later stages biserial, eventually somewhat twisted. Wall finely agglutinated, smooth, finely perforate. Sutures indistinct in early portion, distinct and depressed in biserial portion. Aperture an elongate opening bordered by a lip in face of ultimate chamber, parallel to suture.

*Remarks.* – The number of chambers and the test size are variable in *K. bradyi*. The agglutinated test is composed of calcareous pieces, which exhibits a slightly roughened surface texture. Earlier *Eggerella bradyi* (Cushman) was considered to be a juvenile stage of this species (Parr 1950; Phleger *et al.* 1953). Phleger *et al.* (1953) also suggested that these two species had the same bathymetric range. Pflum & Frerichs (1976) also suggested that there may be an intergradational series between *K. bradyi* and *E. bradyi*.

*Ecology.* – According to Brady (1884), and Pflum & Frerichs (1976) *Karreriella bradyi* has its upper depth limit in the lowermost neritic zones, which is somewhat similar to that of *Eggerella bradyi*. In the South China Sea, *K. bradyi* has been found midway between the lysocline (3,200 m) and CCD (3,800 m), and extends to depth below the CCD (Miao & Thunell 1993).

*Distribution at Sites 502A and 503B. –* *Karreriella*   *bradyi* is a very rare species at both Site 502A and Site 503B. The species is more abundant at Site 502A than at Site 503B, where it occurs only at one depth [503B-14-1, 128–130 cm].

# Subfamily Valvulininae Berthelin, 1880

# Genus *Martinottiella* Cushman, 1933

## *Martinottiella communis* (d'Orbigny, 1826)

Fig. 13H

*Synonymy. –* □1826 *Clavulina communis* d'Orbigny, p. 268, no. 4. □1933a *Martinottiella communis* (d'Orbigny) – Cushman, p. 37, Pl. 4:6–8. □1933c *Martinottiella communis* (d'Orbigny) – Cushman, p. 122, Pl. 12:11. □1937b *Listerella communis* (d'Orbigny) – Cushman, p. 148, Pl. 17:4–9. □1942 *Schenckiella communis* – Thalmann, p. 463. □1950 *Martinottiella communis* (d'Orbigny) – Asano, p. 3, Pl. 3:16–17. □1960 *Martinottiella communis* (d'Orbigny) – Barker, p. 98, Pl. 48:3–4; 6–8. □1964 *Schenckiella communis* n.sp. – Thalmann, 1942 – Leroy, p. 19, Pl. 1:17. □1974 *Martinottiella occidentalis* (Cushman) – Leroy & Levinson, p. 6, Pl. 1:18 (not *Clavulina occidentalis*, Cushman, 1922a). □1980 *Martinottiella communis* (d'Orbigny) – Ingle *et al.* p. 140, Pl. 4:14–15. □1980 *Martinottiella communis* (d'Orbigny) – Keller, Pl. 1:12. □1980 *Martinottiella communis* (d'Orbigny) – Thompson, Pl. 8:9. □1981 *Martinottiella communis* (d'Orbigny) – Cole, pp. 45–46, Pl. 17:24. □1984b *Martinotiella communis* (d'Orbigny) [*sic*] – Boersma, Pl. 1:2–3. □1985 *Martinottiella communis* (d'Orbigny) – Kohl, p. 33, Pl. 4:2. □1986 *Martinottiella communis* (d'Orbigny) – Boersma, Pl. 3:5. □1989 *Martinottiella communis* (d'Orbigny) – Hermelin, p. 34, Pl. 2:5–6. □1990 *Martinottiella communis* (d'Orbigny) – Thomas *et al.*, Pl. 1:11.

*Description. –* Test free, elongate; initial portion trochospiral, reduced to triserial and later uniserial; chambers indistinct in early portion, more distinct in the uniserial portion. Sutures slightly depressed in uniserial stage. Wall finely agglutinated, smooth, finely perforate. Aperture a small round opening or slightly elliptical at the center of the ultimate chamber.

*Distribution at Sites 502A and 503B. – Martinottiella communis* is a rare species with scattered occurrences.

## *Martinottiella milletti* (Cushman, 1936)

*Synonymy. –* □1936a *Listerella milletti* – Cushman, p. 41, Pl. 6:10. □1988 *Martinottiella milletti* (Cushman) – Zheng, p. 106, Pls. 49:9–10; 50:1. □1990 *Martinottiella milletti* (Cushman) – Ujiié, p. 14, Pl. 1:9.

*Description. –* Test free, elongate, triserial in early stages, later uniserial; chambers indistinct in early portion, more distinct in the uniserial portion. Sutures indistinct, slightly depressed. Wall coarsely agglutinated. Aperture terminal, rounded.

*Remarks. –* One of the most characteristic features of this species is its rough surface. *Martinottiella milletti* has a coarser surface than *Martinottiella communis.*

*Distribution at Sites 502A and 503B. – Martinottiella milletti* is represented by a single specimen at Site 502A at 134.28 m. It is absent at Site 503B.

# Family Miliolidae Ehrenberg, 1839

# Genus *Nummoloculina* Steinmann

## *Nummoloculina irregularis* (d'Orbigny, 1839)

Fig. 13I

*Synonymy. –* □1839a *Biloculina irregularis* n.sp. – d'Orbigny, p. 7, Pl. 8:20–21. □1953 *Nummoloculina irregularis* (d'Orbigny) – Phleger *et al.* p. 28, Pl. 5:19–20. □1964 *Nummoloculina irregularis* (d'Orbigny) – Smith, p. 28. □1978 *Nummoloculina irregularis* (d'Orbigny) – Lohmann, p. 25, Pl. 2:16–17.

*Description. –* Test free, nearly as long as wide; chambers quinqueloculine, not inflated, periphery broadly rounded. Sutures distinct. Wall calcareous, porcellaneous, smooth, polished. Aperture elliptical, with a long broad tooth.

*Ecology. –* Lohmann (1978) found that this species decreases with decreasing temperature and increasing alkalinity, depth, and silica in the western south Atlantic Ocean. *Nummoloculina* is considered as epifaunal (Corliss 1985, 1991; Corliss & Chen 1988). In the Sulu Sea, it is found restricted to the top 1 cm (Rathburn & Corliss 1994).

*Distribution at Sites 502A and 503B. – Nummoloculina irregularis* at Site 502A is represented by only one specimen at 151.70 m [502A-40-2, 1–3 cm]. Only a few specimens occur at Site 503B.

# Subfamily Quinqueloculininae Cushman, 1917

## Genus *Quinqueloculina* d'Orbigny, 1826

## *Quinqueloculina weaveri* Rau, 1948

Fig. 13J

*Synonymy.* – □1948 *Quinqueloculina weaveri* n.sp. – Rau, 1948, pp. 159–160, Pl. 28:1–3. □1953 *Quinqueloculina* cf. *weaveri* (Rau) – Phleger *et al.* p. 28, Pl. 5:13–14. □1978 *Quinqueloculina* d'Orbigny – Lohmann, p. 25, Pl. 4:17–18. □1979 *Quinqueloculina* cf. *weaveri* (Rau) – Corliss, p. 6, Pl. 1:12–14. □1980 *Quinqueloculina weaveri* Rau, 1948 – Boltovskoy, 1980, Pl. 3:4a–b. □1981 *Sigmoilina edwardi* (Schlumberger) – Resig [not *Planispirina* (*Sigmoilina*) *edwardi* Schlumberger, 1887], Pl. 5:10. □1985 *Quinqueloculina* cf. *weaveri* (Rau) – Thomas, Pl. 1:11. □1986 *Sigmoilina edwardi* (Schlumberger) – Kurihara & Kennett, p. 1069, Pl. 1:16–17. □1988 *Sigmoilina edwardi* (Schlumberger) – Kurihara & Kennett, Pl. 1:4–7. □1990 *Quinqueloculina weaveri* Rau, 1948 – Ujiié, p. 14, Pl. 3:1a–b; 2a–b.

*Description.* – Test free, fusiform, nearly as long as wide; chambers prominently triangular in cross-section. Sutures distinct, slightly depressed. Wall calcareous, porcellaneous, smooth, polished. Aperture elliptical, occasionally with a faint tooth.

*Remarks.* – Rau (1948) and Phleger *et al.* (1953) described this species as toothless throughout its lifecycle. However, among the many species analysed here, a few specimens of *Quinqueloculina weaveri* with a faint tooth were found. Boltovskoy (1978) reported that this species is subject to considerable morphological variations in its general outline, character of chambers and size. The aperture is also variable.

*Ecology.* – Corliss (1979) found *Q. weaveri* from 2,500 to 4,500 m and at 67% of the investigated stations with species frequency values between 1 and 5% in the southeast Indian Ocean. In the Sulu Sea, *Q. weaveri* has been found dominant together with *Pyrgo murrhina* between 1,400 and 2,200 m water depth, whereas neither *Quinqueloculina* nor *Pyrgo* contribute significantly to the benthic fauna in the South China Sea (Miao & Thunell 1993). In the South China Sea, Miao & Thunell (1996) found *Q. weaveri* relatively abundant in the glacial fauna at 3,500 m water depth.

*Distribution at Sites 502A and 503B.* – *Quinqueloculina weaveri* is a common species in both sites, with absolute abundances between 0 and 9 at Site 502A and between 0 and 7 at Site 503B.

## *Quinqueloculina venusta* Karrer, 1868

Fig. 13K–L

*Synonymy.* – □1868 *Quinqueloculina venusta* n.sp. – Karrer, p. 147, Pl. 2:6. □1884 *Miliolina venusta* (Karrer) – Brady (part), p. 162, Pl. 5:5a–c (not 7a–c). □1917a *Quinqueloculina venusta* Karrer, 1868 – Cushman, pp. 45–46, Pl. 11:1a–c. □1921 *Quinqueloculina venusta* Karrer, 1868 – Cushman, pp. 420–421, Pl. 91:2a–c. □1953 *Quinqueloculina venusta* Karrer, 1868 – Phleger *et al.*, p. 27, Pl. 5:11–12. □1960 *Quinqueloculina venusta* Karrer? – Barker, p. 10, Pl. 5:5a–c. □1978 *Quinqueloculina venusta* Karrer, 1868 – Boltovskoy, Pl. 6:32–33. □1978 *Quinqueloculina venusta* Karrer, 1868 – Lohmann, p. 25, Pl. 4:8–9. □1979 *Quinqueloculina venusta* Karrer, 1868 – Corliss, p. 6, Pl. 1:9–11. □1981 *Triloculina* sp. A – Burke, Pl. 1:7; 10–11. □1986 *Quinqueloculina venusta* Karrer, 1868 – Kurihara & Kennett, Pl. 1:14–15. □1989 *Quinqueloculina venusta* Karrer, 1868 – Hermelin, p. 36, Pl. 2:11, 14. □1990 *Quinqueloculina venusta* Karrer, 1868 – Ujiié, p. 15, Pl. 3:3a–b; 4a–b.

*Description.* – Test free, fusiform, 1–1.5 times as long as wide, prominently triangular in cross-section, peripheral angles bluntly angular. Sutures distinct, slightly depressed. Wall calcareous, porcellaneous, smooth. Aperture circular with a thickened lip and a short tooth.

*Remarks.* – Boltovskoy (1978) found that specimens from the Indian Ocean had, on the average, less sharp peripheral angles, and in many cases the edge of the test was less elongated.

*Ecology.* – Brady (1884) found *Q. venusta* at 14 stations, of which 12 were in water depth between 3,290 and 4,936 m. In the southeast Indian Ocean, Corliss (1979) reported *Quinqueloculina venusta* from the lower bathyal zone. At the Ontong–Java Plateau, Burke (1981) found *Q. venusta* (as *Triloculina* sp. A) in the abyssal depth, whereas Hermelin (1989) reported this species as a rare one in scattered samples in DSDP Hole 586A (waterdepth 2,207 m) from the Ontong–Java Plateau.

*Distribution at Sites 502A and 503B.* – *Quinqueloculina venusta* is a relatively common species at Site 502A, but with low abundances between 0 and 9 individuals per sample. It is a rare species with scattered occurrences at Site 503B.

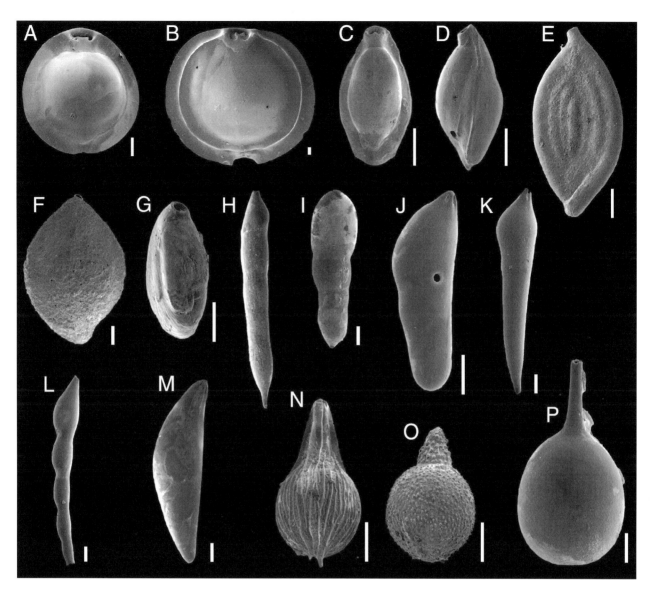

*Fig. 14.* Scale bar 100 µm. □A. *Pyrgo depressa* (d'Orbigny), Side view, Hole 502A, 89.58 m. □B. *Pyrgo murrhina* (Schwager), Side view, Hole 502A, 47.76 m. □C, D. *Pyrgo oblonga* (d'Orbigny); C, Side view, Hole 502A, 65.51 m; D, Edge view, Hole 502A, 5043 m. □E. *Sigmoilina tenuis* (Czjzek), Side view, Hole 502A, 105.58 m. □F. *Sigmoilopsis schlumbergeri* (Silvestri), Side view, Hole 502A, 100.73 m. □G. *Triloculina tricarinata* d'Orbigny, Side view, Hole 502A, 84.96 m. □H. *Chrysalogonium lanceolum* Cushman & Jarvis, Side view, Hole 503B. 74.82 m. □I. *Chrysalogonium tenuicostatum* Cushman & Bermúdez, Side view, Hole 503B, 67.26 m. □J. *Dentalina communis* d'Orbigny, Side view, Hole 502A, 97.80 m. □K. *Dentalina* cf. *communis* Hermelin, Side view, Hole 502A, 80.46 m. □L. *Dentalina filiformis* (Reuss), Side view, Hole 502A, 47.76 m. □M. *Dentalina intorta* (Dervieux), Side view, Hole 502A, 97.80 m. □N. *Lagena advena* Cushman, Side view, Hole 502A, 113.42. □O. *Lagena hispida* Reuss, Side view, Hole 502A, 104.54 m. □P. *Lagena hispidula* Cushman, Side view, Hole 502A, 66.51 m.

## Genus *Pyrgo* Defrance, *in* Blainville, 1824

### *Pyrgo depressa* (d'Orbigny, 1826)

Fig. 14A

*Synonymy.* – □1826 *Biloculina depressa* n.sp. – d'Orbigny, p. 98. □1884 *Biloculina depressa* d'Orbigny, 1826 – Brady, p. 146, Pl. 2:13–15. □1929a *Pyrgo depressa* (d'Orbigny) – Cushman, p. 71, Pl. 19:4–5. □1960 *Pyrgo depressa* (d'Orbigny) – Barker, p. 4, Pl. 2:12;16–17. □1964 *Pyrgo depressa* (d'Orbigny) – Leroy, p. F21, Pl. 12:29–30. □1978

*Pyrgo depressa* (d'Orbigny) – Boltovskoy, p. 167, Pl. 6:25. □1978 *Pyrgo depressa* (d'Orbigny) – Wright, p. 716. □1990 *Pyrgo depressa* (d'Orbigny) – Ujiié, p. 16, Pl. 4:1a–b; 2a–b.

*Description.* – Test free, compressed, circular in side view; periphery carinate, plate-like, increasing in width relative to chamber size. Test biloculine; chambers inflated. Sutures depressed. Wall calcareous, porcellaneous, smooth, ornamented with a broad keel that extends completely around the periphery. Aperture terminal, elongate slit with thin lips.

*Remarks.* – This species is very similar to *P. murrhina*, and many workers report it as a *P. murrhina*. However, the aperture of this species is very different, since it has a very elongate, narrow slit near the apex of the test.

*Distribution at Sites 502A and 503B.* – *Pyrgo depressa* is a rare species with scattered occurrences at both Site 502A and Site 503B.


## *Pyrgo murrhina* (Schwager, 1866)

Fig. 14B

*Synonymy.* – □1866 *Biloculina murrhina* n.sp. – Schwager, p. 203, Pl. 4:15. □1884 *Biloculina depressa*, var. *murrhyna* Schwager – Brady, p. 146, Pl. 2:10–11, 15. □1929a *Pyrgo murrhina* (Schwager) – Cushman, p. 71, Pl. 19:6a–b. □1932a *Pyrgo murrhina* (Schwager) – Cushman, pp. 64–65, Pl. 15:1–3. □1951 *Pyrgo murrhina* (Schwager) – Phleger & Parker, p. 7, Pl. 3:11. □1953 *Pyrgo murrhina* (Schwager) – Phleger *et al.*, pp. 28–29, Pl. 5:22–24. 1960 *Pyrgo murrhyna* (Schwager) – Barker, p. 4, Pl. 2:10–11; 15. □1964 *Pyrgo murrhina* (Schwager) – Akers & Dorman, p. 49, Pl. 3:14–15. □1964 *Pyrgo murrhina* (Schwager) – Leroy, p. 21, Pl. 12:32–33. □1971 *Pyrgo murrhina* (Schwager) – Bock, p. 24, Pl. 8:14. □1974 *Pyrgo murrhina* (Schwager) – Leroy & Levinson, p. 6, Pl. 2:5. □1978 *Pyrgo murrhina* (Schwager) – Boltovskoy, Pl. 6:26. □1979 *Pyrgo murrhina* (Schwager) – Corliss, p. 6, Pl. 1:15–18. □1981 *Pyrgo murrhyna* (Schwager) – Burke, Pl. 1:9. □1981 *Pyrgo murrhyna* (Schwager) – Cole, pp. 52–53, Pl. 8:9. □1981 *Pyrgo murrhyna* (Schwager) – Resig, Pl. 5:9. □1984a *Pyrgo murrhina* (Schwager) – Boersma, Pl. 5:2, Pl. 7:2. □1984 *Pyrgo murrhina* (Schwager) – Murray, Pl. 3:3. □1985 *Pyrgo murrhina* (Schwager) – Kohl, p. 35, Pl. 5:1. □1985 *Pyrgo murrhina* (Schwager) – Thomas, Pl. 1:10. □1986 *Pyrgo murrhina* (Schwager) – Kurihara & Kennett, Pl. 1:13. □1986 *Pyrgo murrhina* (Schwager) – Morkhoven *et al.* pp. 50–52, Pl. 15:1–2. □1989 *Pyrgo murrhina* (Schwager) – Hermelin, p. 36, pl:15–16. □1990 *Pyrgo murrhyna* (Schwager) – Thomas *et al.*, Pl. 8:3. □1990 *Pyrgo murrhina* (Schwager) – Ujiié, p. 16, Pl. 4:3a–b; 4a–b; 5a–b.

*Description.* – Test free, inflated, subcircular inside view, subovate in transverse section with margin extended and carinate. Test biloculine; chambers inflated. Wall calcareous, porcellaneous, smooth, ornamented with a broad keel that extends completely about the periphery, but possessing a small sinus at posterior end. Aperture terminal, near junction of last two chambers, with a prominent tubular neck, round to ovate, with a bifid tooth.

*Remarks.* – *Pyrgo murrhina* varies a great deal in size and shape. The aperture of the juveniles is much less broad than that of mature individuals of the species and the characteristic sinus in the carina is often lacking in adult specimens (Phleger *et al.*, 1953).

*Ecology.* – *Pyrgo murrhina* is generally found in the lower middle to lower bathyal zone (Kennett 1982). In the western equatorial Pacific, *Pyrgo* is found in high abundance along with *Quinqueloculina* and *Uvigerina* between 1,600 and 2,400 m water depth, within the deep oxygen minimum layer (Culp 1977). Corliss (1979) reported *P. murrhina* as the most abundant *Pyrgo* in the deep sea of the southeast Indian Ocean, and it occurs between 2,500 and 4,600 m. In the Sulu Sea, *P. murrhina* is common between 1,400 and 2,200 m (Miao & Thunell 1993). Miao & Thunell (1996) found *P. murrhina* abundant during both the Holocene and glacial stage 2, with a minimum value at the stage 1–2 boundary. On the Ontong Java Plateau, *P. murrhina* has its maximum abundances during glacial periods (Burke *et al.* 1993). *Pyrgo murrhina* is reported to be more abundant in areas with high productivity in the surface waters (Boersma 1985; Woodruff & Savin 1989).

*Distribution at Sites 502A and 503B. Pyrgo murrhina* is an abundant species throughout the studied interval of Site 502A with absolute abundances between 0 and 142 (Appendix), while the species exhibits a more scattered distribution at Site 503B.


## *Pyrgo oblonga* (d'Orbigny, 1839)

Fig. 14C–D

*Synonymy.* – □1839a *Biloculina oblonga* n.sp. – d'Orbigny, p. 163, Pl. 8:21–23. □1953 *Pyrgo oblonga* (d'Orbigny) – Phleger *et al.*, p. 29, Pl. 5:25–26. □1981 *Pyrgo* cf. *oblongus* (d'Orbigny) – McCulloch, p. 60, Pl. 20:9. □1990 *Pyrgo oblonga* (d'Orbigny) – Ujiié, p. 16, Pl. 3:9a–b.

*Description.* – Test free, elongate, width less than two-thirds of length. Test biloculine; chambers distinct, inflated. Sutures depressed, partly concealed by acute, peripheral margin of preceding chamber extending laterally, basally, anteriorly. Wall calcareous, smooth, white, thin, polished, imperforate. Aperture subterminal, extended over opening of last two chambers, small, ovate, oblique, with a bifid tooth.

*Remarks.* – *Pyrgo oblonga* resembles *P. murrhina*, but is smaller and has a more elongate test. The prolongation of the neck and the angle of the periphery vary in this species (Phleger *et al.* 1953).

*Distribution at Sites 502A and 503B.* – *Pyrgo oblonga* occurs in many samples but in relatively low abundances at both Site 502A and Site 503B.

# Genus *Sigmoilina* Schlumberger 1887

## *Sigmoilina tenuis* (Czjzek, 1848)

Fig. 14E

*Synonymy.* – □1945 *Sigmoilina tenuis* (Czjzek) – Cushman & Todd, p. 10, Pl. 2:4. □1946 *Sigmoilina tenuis* (Czjzek) – Cushman & Gray, p. 5, Pl. 1:14–16. □1951 *Sigmoilina tenuis* (Czjzek) – Marks, p. 39, Pl. 5:7. □1951 *Sigmoilina tenuis* (Czjzek) – Phleger & Parker, p. 8, Pl. 4:7. □1953 *Sigmoilina tenuis* (Czjzek) – Phleger *et al.*, p. 28, Pl. 5:18. □1957 *Sigmoilina tenuis* (Czjzek) – Todd & Brönnimann, p. 29, Pl. 4:3. □1958 *Sigmoilina tenuis* (Czjzek) – Parker, p. 257, Pl. 1:24. □1964 *Sigmoilina tenuis* (Czjzek) – Leroy, p. 20, Pl. 16:32–33. □1978 *Sigmoilina tenuis* (Czjzek) – Wright, p. 717, Pl. 7:18. □1985 *Sigmoilinita tenuis* (Czjzek) Kohl, p. 36, Pl. 5:5.

*Description.* – Test free, compressed, small, ovate in outline, periphery rounded. Chambers distinct, inflated, slightly more than 180° from one other, forming a sigmoidal curve in cross-section. Sutures distinct, depressed. Wall smooth, calcareous. Aperture terminal, circular, on a short neck.

*Distribution at Sites 502A and 503B.* – *Sigmoilina tenuis* is relatively common at Site 502A above 111.48 m (3.95 Ma). It is a rare species with few occurrences at Site 503B.

# Genus *Sigmoilopsis* Finlay, 1947

## *Sigmoilopsis schlumbergeri* (Silvestri, 1904)

Fig. 14F

*Synonymy.* – □1904a *Sigmoilina schlumbergeri* – Silvestri, p. 267, 269, Pl. 7:12-14, pp. 481-482, Text-figs. 6-7. □1945 *Sigmoilina schlumbergeri* Silvestri, 1904a – Cushman & Todd, p. 11, Pl. 2:3. □1947 *Sigmoilopsis schlumbergeri* (Sylvestri) – Finlay, p. 270. □1953 *Sigmoilina schlumber-geri* Silvestri, 1904a – Phleger *et al.*, p. 28, Pl. 5:17. □1960 *Sigmoilopsis schlumbergeri* (Sylvestri) – Barker, p. 16, Pl. 8:1–4. □1971 *Sigmoilina schlumbergeri* Silvestri, 1904a – Bock, p. 25, Pl. 9:1-2. □1971 *Sigmoilopsis schlumbergeri* (Sylvestri) – Bock, p. 25, Pl. 9:1–2. □1978 *Sigmoilina schlumbergeri* Silvestri, 1904a – Boltovskoy, Pl. 7:5-6. □1981 *Sigmoilopsis schlumbergeri* (Sylvestri) – Cole, p. 55, Pl. 10:1. □1984b *Sigmoilina schlumbergeri* Silvestri, 1904a – Boersma, Pl. 1:8. ]1984a *Sigmoilopsis schlumbergeri* (Sylvestri) – Boersma, Pl. 8:2. □1985 *Sigmoilopsis schlumbergeri* (Sylvestri) – Hermelin & Scott, p. 217, Pl. 2:6. □1985 *Sigmoilopsis schlumbergeri* (Sylvestri) – Kohl, p. 36, Pl. 5:6. □1986 *Sigmoilopsis schlumbergeri* (Sylvestri) – Belanger & Berggren, p. 331, Pl. 1:5a-b. □1986 *Sigmoilopsis schlumbergeri* (Sylvestri) – Boersma, Pl. 14:1–2. □1986 *Sigmoilopsis schlumbergeri* (Sylvestri) – Kurihara & Kennett, Pl. 1:11–12. □1986 *Sigmoilopsis schlumbergeri* (Sylvestri) – van Morkhoven *et al.*, pp. 57-59, Pl. 18:1a–e. □1989 *Sigmoilopsis schlumbergeri* (Sylvestri) – Hermelin, p. 38. □1990 *Sigmoilopsis schlumbergeri* (Sylvestri) – Thomas *et al.*, Pl. 8:4. □1990 *Sigmoilopsis schlumbergeri* (Sylvestri) – Ujiié, p. 16, Pl. 3:10a–b.

*Description.* – Test free, elongate, ovate in side view, sub-triangular in transverse section. Chambers indistinct, inflated, slightly greater than 180° from one other, forming a sigmoidal curve seen in transverse section. Sutures indistinct. Wall finely agglutinated. Aperture terminal, rounded, with a bifid tooth, and a lip.

*Ecology.* – *Sigmoilopsis schlumbergeri* occurs from the upper neritic zone down to the upper middle bathyal zone in the Gulf of Mexico (Phleger 1951; Pflum & Frerichs 1976), although it is found in greatest abundance in the upper part of the upper middle bathyal zone (Phleger 1951).

*Distribution at Sites 502A and 503B.* – *Sigmoilopsis schlumbergeri* is an abundant species throughout the studied interval of Site 502A with absolute abundances between 0 and 31. It is absent at Site 503B.

# Genus *Triloculina* d'Orbigny, 1826

## *Triloculina tricarinata* d'Orbigny, 1826

Fig. 14G

*Synonymy.* – □1826 *Triloculina tricarinata* n.sp. – d'Orbigny, p. 299, no. 7, mod. no. 94. □1953 *Triloculina tricarinata* d'Orbigny, 1826 – Phleger *et al.*, p. 28, Pl. 5:21.

□1954 *Triloculina tricarinata* d'Orbigny, 1826 – Cushman *et al.*, p. 340, Pl. 85:15–16. □1960 *Triloculina tricarinata* d'Orbigny, 1826 – Barker, p. 6, Pl. 3:17a–b. □1964 *Triloculina tricarinata* d'Orbigny, 1826 – Akers & Dorman, p. 57, Pl. 3:21. □1964 *Triloculina tricarinata* d'Orbigny, 1826 – Feyling–Hanssen, p. 258, Pl. 6:7–8. □1964 *Triloculina tricarinata* d'Orbigny, 1826 – Leroy, p. 20, Pl. 3:32–33. □1971 *Triloculina tricarinata* d'Orbigny, 1826 – Bock, p. 28, Pl. 12:1–2. □1971 *Triloculina tricarinata* d'Orbigny, 1826 – Schnitker, p. 212, Pl. 3:10. □1981 *Triloculina tricarinata* d'Orbigny, 1826 – Cole, p. 55, Pl. 10:2. □1985 *Triloculina tricarinata* d'Orbigny, 1826 - Hermelin & Scott, p. 218, Pl. 2:7. □1989 *Triloculina tricarinata* d'Orbigny, 1826 – Hermelin, p. 38, Pl. 3:6–7. □1990 *Triloculina tricarinata* d'Orbigny, 1826 – Thomas *et al.*, Pl. 8:5. □1991 *Triloculina tricarinata* d'Orbigny, 1826 – Scott & Vilks, Pl. 3:3–4.

*Description.* – Test free, strongly triangular in apertural view, slightly longer than broad, sides straight or slightly concave, three chambers visible. Sutures distinct, slightly depressed. Wall calcareous, surface smooth. Aperture circular, with a small bifid tooth.

*Ecology.* – The genus *Triloculina* is generally related to low-oxygen conditions (Burke 1981).

*Distribution at Sites 502A and 503B.* – *Triloculina tricarinata* is a very rare species with few occurrences at Site 502A. It is absent at Site 503B.

# Family Peneroplidae Schultze, 1854

# Genus *Monalysidium* Chapman, 1900

## *Monalysidium politum* Chapman, 1900

*Synonymy.* – □1900 *Monalysidium politum* n.sp. – Chapman, p. 3. □1960 *Monalysidium politum* Chapman – Barker, p. 26, Pl. 13:24–25. □1977 *Monalysidium* (?) cf. *politum* Chapman – McCulloch, p. 232, Pl. 100:15.

*Description.* – Test free, elongate, uniserial. Chambers about uniform in length, basal chamber short. Sutures distinct, depressed. Wall calcareous, semitransparent, finely perforate between longitudinal, beaded costae with punctate, evenly spaced costae interrupted by the sutures. Aperture terminal, at the center of the ultimate chamber with a small, short tubular neck and a round opening.

*Distribution at Sites 502A and 503B.* – *Monalysidium politum* is absent at Site 502A and is represented by a single specimen at Site 503B (3.13 m).

# Suborder *Rotaliina* Delage & Hérouard, 1896

# Superfamily Nodosariacea Ehrenberg, 1838

# Family Nodosariidae Ehrenberg, 1838

# Subfamily Nodosariinae Ehrenberg, 1838

# Genus *Chrysalogonium* Schubert, 1907

## *Chrysalogonium lanceolum* Cushman & Jarvis, 1934

Fig. 14H

*Synonymy.* – □1934 *Chrysalogonium lanceolum* n.sp. – Cushman & Jarvis, p. 75, Pl. 10:16. □1989 *Chrysalogonium lanceolum* Cushman & Jarvis, 1934 – Hermelin, p. 9, Pl. 3:12–13.

*Description.* – Test free, very elongate, gradually tapering, widest close to the apertural end. Chambers numerous, gradually increasing in length as added. Sutures distinct, slightly depressed. Wall calcareous, smooth. Aperture terminal, cone-shaped, a series of rounded or slightly elongate pores.

*Distribution at Sites 502A and 503B.* – This species is absent at Site 502A and has scattered occurrences at Site 503B.

## *Chrysalogonium longicostatum* Cushman & Jarvis, 1934

*Synonymy.* – □1934 *Chrysalogonium longicostatum* – Cushman & Jarvis, p. 74, Pl. 10:12a–b. □1989 *Chrysalogonium tenuicostatum* Cushman & Jarvis, 1934 – Hermelin, pp. 39–40, Pl. 3:14–15.

*Description.* – Test free, elongate; chambers distinct, sutures slightly depressed. Wall calcareous, ornamented with longitudinal costae covering nearly the entire test, the striation slightly twisted. Aperture terminal, consisting of a definite sieve plate composed of concentric rings of rounded or slightly polygonal openings.

*Distribution at Sites 502A and 503B.* – *Chrysalogonium longicostatum* is a very rare species at both Site 502A and Site 503B.

## *Chrysalogonium tenuicostatum* Cushman & Bermúdez, 1936

Fig. 14I

*Synonymy.* – □1936 *Chrysalogonium tenuicostatum* n.sp. – Cushman & Bermúdez, pp. 27–28, Pl. 5:3–5. □1989 *Chrysalogonium tenuicostatum* Cushman & Bermúdez, 1936 – Hermelin, p. 40, Pl. 3:16–18. □1986 *Chrysalogonium equisetiformis* (Schwager) – Boersma, Pl. 9:1–3 (not *Nodosaria equisetiformis* Schwager, 1866).

*Description.* – Test free, elongate, very slightly tapering, straight. Chambers increasing gradually in length as added, sides slightly convex. Sutures limbate, slightly depressed. Wall calcareous, ornamented with fine longitudinal costae covering nearly the entire test, slightly spiral. Aperture terminal, consisting of a very fine circular sieve plate with many openings.

*Remarks.* – *Chrysalogonium tenuicostatum* has much coarser apertural openings and finer costae than *Chrysalogonium longicostatum*.

*Distribution at Sites 502A and 503B.* – *Chrysalogonium longicostatum* is a very rare species at both Site 502A and Site 503B.

## Genus *Dentalina* Risso, 1826

## *Dentalina communis* d'Orbigny, 1826

Fig. 14J

*Synonymy.* – □1826 *Nodosaria (Dentalina) communis* – d'Orbigny, p. 254. □1840 *Dentalina communis* sp.nov. d'Orbigny, – d'Orbigny, p. 13, Pl. 1:4. □1960 *Dentalina communis* d'Orbigny, 1840 – Barker, p. 130, Pl. 2:21–22. □1964 *Dentalina communis* d'Orbigny, 1840 □1964 *Dentalina communis* d'Orbigny, 1840 – Akers & Dorman, p. 32, Pl. 6:12. □1964 *Dentalina communis* d'Orbigny, 1840 – Leroy, p. 23, Pl. 15:28. □1966 *Dentalina communis* d'Orbigny, 1840 – Todd, p. 26, Pl. 12:1. □1971 *Dentalina communis* d'Orbigny, 1840 – Bock, pp. 38–39, Pl. 14:16–17. □1971 *Dentalina communis* d'Orbigny, 1840 – Schnitker, p. 196, Pl. 3:20a–b. □1978 *Dentalina communis* d'Orbigny, 1840 – Boltovskoy, Pl. 3:23. □1985 *Dentalina communis* d'Orbigny, 1840– Kohl, p. 39, Pl. 7:3–5. □1989 *Dentalina communis* d'Orbigny, 1840 – Hermelin, p. 40, Pl. 4:1–2. □1990 *Dentalina communis* d'Orbigny, 1840 – Ujiié, p. 17, Pl. 4:8a–b, 9.

*Description.* – Test free, uniserial, slightly curved, rounded in transverse section; chambers in the adult, increasing gradually in size, early chambers subglobular, later chambers inflated. Sutures oblique, depressed in later half of test. Wall calcareous, hyaline, smooth, finely perforate, often ornamented with an initial spine. Aperture terminal, radiate.

*Distribution at Sites 502A and 503B.* – *Dentalina communis* is the most common species of this genus at Site 502A and appears at most levels in relatively low abundances (0–4 individuals per sample). It is a very rare species at Site 503B with scattered occurrences.

## *Dentalina* cf. *communis*

Fig. 14K

*Description.* – Test free, uniserial, slightly curved, compressed laterally. Chambers increasing slightly in size as added. Sutures oblique, mostly flush with surface. Wall calcareous, hyaline, smooth, finely perforate, often ornamented with initial spine. Aperture terminal radiate.

*Remarks.* – *Dentalina* cf. *communis* is similar to *D. communis* and is described by Hermelin (1989).

*Distribution at Sites 502A and 503B.* – *Dentalina* cf. *communis* is a very rare species with few occurrences at both investigated sites.

## *Dentalina filiformis* (Reuss, 1826)

Fig. 14L

*Synonymy.* – □1826 *Nodosaria filiformis* – d'Orbigny, p. 253, Pl. 15:2. □1845 *Nodosaria (Dentalina) filiformis* sp.nov. – Reuss, p. 28, Pl. 12:28. □1871 *Dentalina filiformis* (Reuss) – Parker, Jones & Brady, p. 156, Pl. 9:28. □1960 *Dentalina filiformis* (Reuss) – Barker, p. 132, Pl. 63:3–5. □1986 *Dentalina filiformis* (Reuss) – Belanger & Berggren, p. 331.

*Description.* – Test free, elongate, uniserial, slightly curved, circular in cross-section. Chambers 11–15 in the adult, elliptical or ovate, increasing gradually in length toward the apertural end. Sutures usually oblique, slightly depressed in later portion of test. Wall calcareous, hyaline, smooth, finely perforate. Aperture terminal, radiate.

*Distribution at Sites 502A and 503B.* – *Dentalina filiformis* is represented by a single specimen at Site 502A (47.76 m). It is absent at Site 503B.

## *Dentalina intorta* (Dervieux, 1894)

Fig. 14M

*Synonymy.* – □1894 *Nodosaria intorta* – Dervieux, p. 610, Pl. 5:32–34. □1960 *Dentalina intorta* (Dervieux) – Barker, p. 132, Pl. 62:27–31. □1978 *Dentalina intorta* (Dervieux)

– Boltovskoy, Pl. 3:25. □1989 *Dentalina intorta* (Dervieux) – Hermelin, p. 40, Pl. 4:4.

*Description.* – Test free, elongate, uniserial, with one side more or less straight and the other curved, widest at the middle, narrowing towards the apertural as well as the initial end, last chamber making up almost half of test. Sutures distinct, flush with surface. Wall calcareous, hyaline, smooth. Aperture terminal, radiate.

*Distribution at Sites 502A and 503B.* – *Dentalina intorta* is a very rare species at Site 502A and absent at Site 503B.

## Genus *Lagena* Walker & Jacob, 1798

*Lagena* is generally characterized to have a unilocular test with various surface ornamentation, and an aperture on an enlongated neck which may have a philaline lip.

## *Lagena advena* Cushman, 1923

Fig. 14N

*Synonymy.* – □1923 *Lagena advena* – Cushman, p. 6, Pl. 1:4. □1960 *Lagena advena* Cushman, 1923 – Barker (part), p. 118, Pl. 57:30 (not 29). □1964 *Lagena advena* Cushman, 1923 – Leroy, p. 26, Pl. 1:1. □1982 *Lagena advena* Cushman, 1923 – Boltovskoy & de Kahn, p. 433, Pl. 8:17–19. □1989 *Lagena advena* Cushman, 1923 – Hermelin, p. 41, Pl. 4:6.

*Description.* – Test free, unilocular, flask-shaped to ovate in side view, circular in cross-section. Wall calcareous, hyaline, body ornamented with numerous longitudinal costae, oriented from the base to the neck, basal part also ornamented with irregularly arranged short spines. Aperture terminal, a round opening at the end of the elongated neck.

*Distribution at Sites 502A and 503B.* – *Lagena advena* is a rare species with scattered occurrences at Site 502A, while the species is very rare with few occurrences at Site 503B.

## *Lagena alticostata* Cushman, 1913

*Synonymy.* – □1913 *Lagena sulcata* (Walker & Jacob) var. *alticostata* – Cushman, 1913, p. 23, Pl. 9:5a–b. □1982 *Lagena sulcata* (Walker & Jacob) var. *alticostata* – Boltovskoy & de Kahn, p. 440, Pl. 11:10–11. □1989 *Lagena alticostata* Cushman – Hermelin, p. 41, Pl. 4:6.

*Description.* – Test free, unilocular, subglobular. Wall calcareous, hyaline, finely perforate, ornamented with a few prominent primary costae orientated from basal end to the neck, between these are finer secondary costae, run-

ning only to the base of the neck. Aperture terminal, rounded, on a neck.

*Distribution at Sites 502A and 503B.* – *Lagena alticostata* is a very rare species with few occurrences at Site 502A. It is absent at Site 503B.

## *Lagena feildeniana* Brady, 1878

*Synonymy.* – □1878 *Lagena feildeniana* n.sp. – Brady, p. 434, Pl. 20:4. □1884 *Lagena feildeniana* Brady, 1878 – Brady, p. 469, Pl. 58:38–39. □1913 *Lagena feildeniana* Brady, 1878 – Cushman, p. 29, Pl. 15:1–2. □1960 *Lagena feildeniana* Brady, 1878 – Barker, p. 120, Pl. 58:38–39. □1989 *Lagena feildeniana* Brady, 1878 – Hermelin, p. 41, Pl. 4:8.

*Description.* – Test free, unilocular, elongate, rounded base, narrowing towards the aperture. Wall calcareous, hyaline, ornamented with numerous fine costae oriented in longitudinal direction, furrows between costae perforated, costae ending in short spines at the base. Aperture terminal.

*Distribution at Sites 502A and 503B.* *Lagena feildeniana* is a very rare species at both investigated sites.

## *Lagena hispida* Reuss, 1863

Fig. 14O

*Synonymy.* – □1863 *Lagena hispida* – Reuss, p. 335, Pl. 6:77–79. □1884 *Lagena hispida* Reuss, 1863 – Brady, p. 459, Pl. 57:1–4. □1913 *Lagena hispida* Reuss, 1863 – Cushman, p. 13, Pl. 4:4–5, Pl. 5:1. □1960 *Lagena hispida* Reuss, 1863 – Barker, p. 116, Pl. 57:1–4. □1989 *Lagena hispida* Reuss, 1863 – Hermelin, p. 42, Pl. 4:9–10.

*Description.* – Test free, unilocular, short neck on a globular body. Wall calcareous, hyaline, finely perforate, body and neck ornamented with short spines, closely set. Aperture terminal, rounded at the end of the neck.

*Remarks.* – *Lagena hispida* has coarser spines and more globular test than *Lagena hispidula*.

*Distribution at Sites 502A and 503B.* – *Lagena hispida* is a rare species with scattered occurrences at Site 502A. It is very rare at Site 503B.

## *Lagena hispidula* Cushman, 1913

Fig. 14P

*Synonymy.* – □1913 *Lagena hispidula* – Cushman, p. 14, Pl. 5:2–3. □1950 *Lagena hispidula* Cushman, 1913 – Cushman & McCulloch, p. 339, Pl. 45:8–10. □1960

*Lagena hispidula* Cushman, 1913 – Barker, p. 114, Pl. 56:12 (not Pl. 56:10–11, 13) (not *Lagena nebulosa* Cushman). □1982 *Lagena hispidula* Cushman, 1913 – Boltovskoy & de Kahn, p. 437, Pl. 9:26–28. □1989 *Lagena hispidula* Cushman, 1913 – Hermelin, p. 42, Pl. 4:11. □1990 *Lagena hispidula* Cushman, 1913 – Thomas *et al.*, Pl. 8:8. □1990 *Lagena hispidula* Cushman, 1913 – Ujiié, p. 18, Pl. 5:3.

*Description.* – Test free, unilocular, circular to ovate in side view, with long apertural neck. Wall calcareous, hyaline, finely perforate, body ornamented with very fine spines, while apertural neck is smooth. Aperture terminal at the end of a long slender neck.

*Remarks.* – *Lagena hispidula* differs from *Lagena hispida* in that it has a smooth flask-shaped body with a long slender neck rather than a rounded short-necked body with coarser spines.

*Distribution at Sites 502A and 503B.* – *Lagena hispidula* is a rare species with few occurrences at both investigated sites.

## *Lagena meridionalis* Wiesner, 1953

*Synonymy.* – □1953 *Lagena meridionalis* Wiesner – Loeblich & Tappan, p. 59. □1960 *Lagena meridionalis* Wiesner – Barker, p. 119, Pl. 58:19. □1964 *Oolina gracilis* Williamson var. *meridionalis* (Wiesner) – Leroy, pp. 26–27, Pl. 13:37. □1982 *Lagena meridionalis* Wiesner – Boltovskoy & de Kahn, p. 437, Pl. 10:3–4. □1989 *Lagena meridionalis* Wiesner – Hermelin, p. 42, Pl. 4:12. □1990 *Lagena meridionalis* Wiesner – Thomas *et al.*, p. 227.

*Description.* – Test free, unilocular, elongate, rounded in cross-section. Wall calcareous, hyaline, finely perforate, ornamented with six to nine costae, which run from the pointed base to the aperture and including the apertural neck. Additional finer and less elevated costae in between the primary costae. Aperture terminal.

*Distribution at Sites 502A and 503B.* – *Lagena meridionalis* is represented by one specimen at Site 502A (88.84 m).

## *Lagena paradoxa* Sidebottom, 1913

Fig. 15A

*Synonymy.* – □1912 *Lagena foleolata* Reuss var. *paradoxa* – Sidebottom, p. 395, Pl. 16:22–23. □1913 *Lagena foleolata* Reuss var. *paradoxa* – Cushman, p. 18, Pl. 15:3a–b. □1982 *Sipholagena paradoxa* (Sidebottom) – Boltovskoy & de Kahn, p. 447, Pl. 15:23–26. □1989 *Lagena paradoxa* Sidebottom – Hermelin, p. 42, Pl. 4:13. □1990 *Lagena paradoxa* Sidebottom – Ujiié, p. 18, Pl. 5:5.

*Description.* – Test free, unilocular, elongate, rounded base, tapering towards the apertural end, with a short neck. Wall calcareous, hyaline, ornamented with longitudinal costae oriented throughout the entire length of the test, fine crossbars between the costae give the test a reticulate pattern. Aperture terminal, at the end of a short neck.

*Distribution at Sites 502A and 503B.* – *Lagena paradoxa* is a rare species with scattered occurrences at Site 502A. It is represented by only one specimen at Site 503B (89.48 m).

## *Lagena striata* (d'Orbigny, 1863)

*Synonymy.* – □1863 *Lagena striata* (d'Orbigny) – Reuss, p. 327, Pl. 3:44–45, Pl. 4:46–47. □1884 *Lagena striata* (d'Orbigny) – Brady (part), p. 460, Pl. 57:22, 24 (not 28,30). □1913 *Lagena striata* (d'Orbigny) – Cushman, p. 19, Pl. 7:4–5. □1960 *Lagena striata* (d'Orbigny) – Barker, p. 118, Pl. 57:22, 24. □1964 *Lagena striata* (d'Orbigny) – Aker & Dorman, p. 40, Pl. 6:11. □1971 *Lagena striata* (d'Orbigny) – Schnitker, p. 204, Pl. 4:3. □1982 *Lagena striata* (d'Orbigny) f. *typica* – Boltovskoy & de Kahn, p. 439, Pl. 10:35, Pl. 11:1. □1989 *Lagena striata* (d'Orbigny) – Hermelin, p. 43, Pl. 4:14. □1990 *Lagena striata* (d'Orbigny) – Thomas *et al.*, Pl. 8:9.

*Description.* – Test free, unilocular, flask-shaped, rounded in cross-section. Wall calcareous, hyaline, finely perforate, body ornamented with numerous fine costae oriented from the base to the base of the neck. Aperture terminal, circular, at the end of a long slender neck.

*Distribution at Sites 502A and 503B.* – *Lagena striata* is a very rare species with few occurrences at both sites.

## *Lagena substriata* Williamson, 1848

*Synonymy.* – □1848 *Lagena substriata* n.sp. – Williamson, p. 15, Pl. 2:12. □1913 *Lagena striata* (d'Orbigny) var. *substriata* Williamson – Cushman, p. 20, Pl. 8:1–3. □1950b *Lagena substriata* Williamson, 1848 – van Voorthuysen, p. 55, Pl. 1:9. □1964 *Lagena striata* (d'Orbigny) var. *substriata* Williamson – Feyling–Hanssen, p. 294, Pl. 12:6. □1989 *Lagena* sp. 2 – Hermelin, p. 43, Pl. 4:17. □1990 *Lagena substriata* Williamson, 1848 – Ujiié, p. 19, Pl. 5:7.

*Description.* – Test free, unilocular, fusiform, widest at base of test, rounded in cross-section, tapering towards the apertural end. Wall calcareous, hyaline, ornamented with several costae. Aperture terminal on a very short neck.

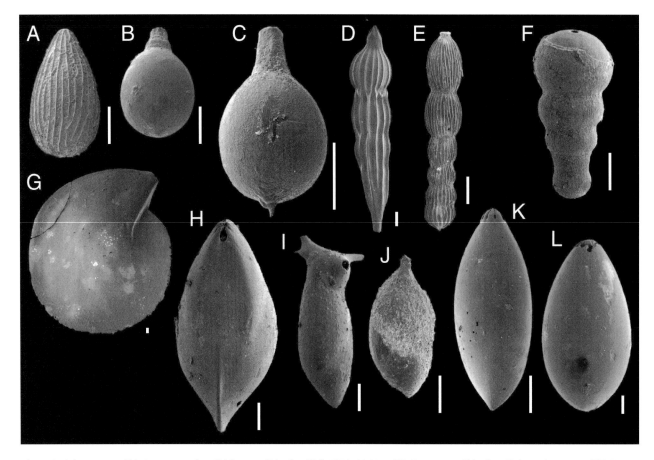

*Fig. 15.* Scale bar 100 µm. □A. *Lagena paradoxa* Sidebottom, Side view, Hole 502A, 83.99 m. □B. *Lagena* sp 1, Side view, Hole 502A, 65.51 m. □C. *Lagena* sp 2, Side view, Hole 502A, 131.38 m. □D. *Nodosaria albatrossi* Cushman, Side view, Hole 502A, 103.63 m. □E. *Orthomorphina challengeriana* (Thalmann), Side view, Hole 502A, 96.82 m. □F. *Orthomorphina* sp 1, Side view, Hole 502A, 66.51 m. □G., H *Lenticulina atlantica* (Barker), G Side view, Hole 502A, 100.73 m; H Edge view, Hole 502A, 105.58 m. □I. *Polymorphinidae formae fistulosae* (Hermelin), Side view, Hole 503B, 56.89 m. □J. *Pyrulina extensa* (Cushman), Side view, Hole 503B, 80.70 m. □K. *Pyrulina fusiformis* (Roemer), Side view, Hole 503B, 62.77 m. □L. *Pyrulina gutta* d'Orbigny, Side view, Hole 502A, 80.46 m.

*Distribution at Sites 502A and 503B. – Lagena substriata* is represented by one specimen at Site 502A (130.30 m). It is absent at Site 503B.

## *Lagena* sp. 1

Fig. 15B

*Description.* – Test free, unilocular, flask-shaped to ovate in side view, circular in cross-section. Wall calcareous, hyaline, finely perforate, body smooth, while aperture neck is ornamented with very fine spines. Aperture terminal, a round opening at the end of the elongated neck.

*Distribution at Sites 502A and 503B. – Lagena* sp. 1 is a rare species with scattered occurrences at Site 502A. It is absent at Site 503B.

## *Lagena* sp. 2

Fig. 15C

*Description.* – Test free, unilocular, flask-shaped to ovate in side view, circular in cross-section. Wall calcareous, hyaline, finely perforate, body smooth, one or two intervening costae terminate below the collar. Aperture terminal, a round opening at the end of the elongated neck.

*Distribution at Sites 502A and 503B. – Lagena* sp. 2 is a rare species with few occurrences at both Site 502A and Site 503B.

## Genus *Nodosaria* Lamarck, 1812

### *Nodosaria albatrossi* Cushman, 1923

Fig. 15D

*Synonymy.* – □1923 *Nodosaria vertebralis* (Batsch) var. *albatrossi* – Cushman, p. 87, Pl. 15:1. □1990 *Nodosaria albatrossi* Cushman – Thomas *et al.*, Pl. 5:1

*Description.* – Test free, elongate, gradually tapering, often slightly curved, proloculus has generally a greater width than those chambers immediately following. Chambers distinct, five to six in the adult, not inflated, except those near apertural end. Sutures distinct, broad, slightly depressed near apertural end. Wall calcareous, hyaline, finely perforate, ornamented by several broad longitudinal costae, generally 15–18 in adult, that continue to the apertural end, initial chamber with a short blunt spine. Aperture slightly extended, radiate.

*Distribution at Sites 502A and 503B.* – *Nodosaria albatrossi* is a rare species with few occurrences at both Site 502A and Site 503B.

### *Nodosaria catesbyi* d'Orbigny, 1839

*Synonymy.* – □1839a *Nodosaria* (*Nodosaria*) *catesbyi* n.sp. – d'Orbigny, in Sagra, p. 16, Pl. 11:8–10. □1872 *Nodosaria proxima* d'Orbigny, 1939a – Silvestri, p. 63, Pl. 6:138–147. □1930 *Nodosaria catesbyi* d'Orbigny, 1939a – Cushman, p. 28, Pl. 5:4. □1957 *Nodosaria catesbyi* d'Orbigny, 1939a – Todd & Brönnimann, p. 31, Pl. 5:4. □1959 *Nodosaria catesbyi* d'Orbigny, 1939a – Boltovskoy, p. 65, Pl. 8:10. □1963 *Nodosaria catesbyi* d'Orbigny, 1939a– Bermúdez & Seiglie, p. 104, Pl. 17:4. □1977 *Lagenonodosaria catesbyi* (d'Orbigny) – LeCalvez, p. 47, p. 48:1–5. □1985 *Nodosaria catesbyi* d'Orbigny, 1939a – Kohl, p. 42, Pl. 6:1.

*Description.* – Test free, uniserial, length slightly greater than twice the width, rounded in cross-section, composed of two to three chambers. Chambers, early subglobular, later pyriform. Sutures distinct, depressed, horizontal. Wall calcareous, hyaline, finely perforate, ornamented with nine to ten distinct costae, which run along the entire test from the base to the apertural end. Aperture terminal, radiate, with a lip at the end of a short neck.

*Distribution at Sites 502A and 503B.* – *Nodosaria catesbyi* is absent at Site 502A and is represented by a single specimen at Site 503B (70.96 m).

## Genus *Orthomorphina* Stainforth, 1952

### *Orthomorphina challengeriana* (Thalmann, 1937)

Fig. 15E

*Synonymy.* – □1937 *Nodogenerina challengeriana* – Thalmann, p. 341. □1952 *Orthomorphina challengeriana* (Thalmann) – Stainforth, p. 8, Text-fig. 1:10. □ 1960 *Orthomorphina challengeriana* (Thalmann) – Barker, p. 136, Pl. 64:25–27. □1964 *Orthomorphina challengeriana* (Thalmann) – Leroy, p. 29, Pl. 15:26. □1978 *Orthomorphina challengeriana* (Thalmann) – Boltovskoy, Pl. 5:16–17. □1989 *Orthomorphina challengeriana* (Thalmann) – Hermelin, p. 44, Pl. 4:20.

*Description.* – Test free, uniseral, straight; chambers inflated. Sutures distinct. Wall calcareous, perforate, ornamented with several longitudinal costae. Aperture terminal, central, rounded, on a short neck, with a rim.

*Distribution at Sites 502A and 503B.* – *Orthomorphina challengeriana* is a rare species with scattered occurrences at Site 502A. At Site 503B it is represented by only one specimen at 56.89 m.

### *Orthomorphina jedlitschkai* (Thalmann, 1937)

*Synonymy.* – □1937 *Nodogenerina jedlitschkai* –Thalmann, p. 341. □1952 *Orthomorphina jedlitschkai* (Thalmann) – Stainforth, pp. 8–9, Text-fig. 1:V. □1960 *Orthomorphina jedlitschkai* (Thalmann) – Barker, p. 130, Pl. 62:1–2. □1989 *Orthomorphina jedlitschkai* (Thalmann) – Hermelin, p. 44.

*Description.* – Test free, uniserial, straight; chamber inflated, generally widest in the middle part of the test. Chambers distinct, sudden changes in the size of the chambers occur. Wall calcareous, smooth, finely perforate. Aperture terminal, rounded.

*Distribution at Sites 502A and 503B.* – *Orthomorphina jedlitschkai* was found in only one sample at Site 502A (72.39 m), but was absent at Site 503B.

### *Orthomorphina* sp. 1

Fig. 15F

*Description.* – Test free, uniserial. Chambers inflated, distinct, sudden changes in the size of the chambers occur. Wall calcareous, finely perforate, ornamented with short spines. Aperture terminal, rounded.

*Distribution at Sites 502A and 503B.* – *Orthomorphina* sp. 1 is a very rare species at Site 502A and is absent at Site 503B.

## Family Vaginulinidae Reuss, 1860

## Subfamily Lenticulininae Chapman, Parr & Collins, 1934

## Genus *Lenticulina* Lamarck, 1804

### *Lenticulina atlantica* (Barker, 1960)

Fig. 15G–H

*Synonymy.* – □1923 *Cristellaria lucida* n.sp. – Cushman, p. 111, Pl. 30:2. □1960 *Robulus atlanticus* nom.nov. – Barker, p. 144, Pl. 69:10–12. □1964 *Lenticulina atlantica* (Barker) – Akers & Dorman, p. 40, Pl. 4:13–14. □1985 *Lenticulina atlantica* (Barker) – Kohl, pp. 46–47, Pl. 10:1–2. □1989 *Lenticulina atlantica* (Barker) – Hermelin, p. 44, Pl. 5:5–6.

*Description.* – Test free, planispiral, involute, circular in side view, biconvex in transverse section, periphery keeled. Sutures slightly curved and depressed, extending to edge of transparent area in umbilicus. Wall calcareous, hyaline, smooth, finely perforate. Aperture radiate, at peripheral angle, with a vertical robuline slit extending downward in the apertural face.

*Ecology.* – In the Sulu Sea, *Lenticulina* generally has its maximum abundance in the 0–1 cm interval and are characterized as epifaunal species (Rathburn & Corliss 1994).

*Distribution at Sites 502A and 503B.* – *Lenticulina atlantica* is well represented at Site 502A and exhibits a more scattered occurrence at Site 503B.

## Family Polymorphinidae d'Orbigny, 1839

## Subfamily Polymorphininae d'Orbigny, 1839

### Polymorphinidae, formae fistulosae

Fig. 15I

*Synonymy.* – □1989 Polymorphinidae, *formae fistulosae* Hermelin, p. 45, Pl. 5:10–11.

*Description.* – Test free, elongate, 'wild-growing' form. Wall calcareous, hyaline, densely ornamented with short spines. Aperture multiple, rounded, on short necks irregularly located on ultimate chamber.

*Distribution at Sites 502A and 503B.* – This species is very rare at both Site 502A and Site 503B with few occurrences.

## Genus *Pyrulina* d'Orbigny, 1839

### *Pyrulina extensa* (Cushman, 1923)

Fig. 15J

*Synonymy.* – □1923 *Polymorphina extensa* sp.nov. – Cushman, p. 156, Pl. 2:104. □1930 *Pyrulina extensa*(Cushman) – Cushman & Ozawa. □1960 *Pyrulina extensa*(Cushman) – Barker, p. 152, Pl. 73:18–19.

*Description.* – Test free, elongate, 1.5–2.5 times as long as wide. Sutures distinct, slightly depressed. Wall calcareous, perforate, ornamented with small spines, especially on the apertural half of the test. Aperture terminal, on a short neck.

*Distribution at Sites 502A and 503B.* – *Pyrulina extensa* is a very rare species with few occurrences at both Site 502A and Site 503B.

### *Pyrulina fusiformis* (Roemer, 1884)

Fig. 15K

*Synonymy.* – □1884 *Polymorphina soraria* var. *cuspidata* n.sp. – Brady, p. 563, Pl. 71:17–19. □1930 *Pyrulina fusiformis* (Roemer) – Cushman & Ozawa, p. 55, Pl. 13:3, 4, 5, 6, 7, 8. □1990 *Pyrulina fusiformis* (Roemer) – Thomas et al., p. 228. □1990 *Pyrulina fusiformis* (Roemer) – Ujiié, p. 21, Pl. 6:12a–b.

*Description.* – Test free, fusiform, almost circular in cross-section. Chambers elongate, arranged in early triserial series, later biserial. Wall calcareous, hyaline. Sutures flush with surface. Aperture radiate.

*Distribution at Sites 502A and 503B.* – *Pyrulina filiformis* is a rare species with scattered occurrences.

### *Pyrulina gutta* d'Orbigny

Fig. 15L

*Synonymy.* – □1826 *Polymorphina (Pyruline) gutta* – d'Orbigny, p. 267. □1989 *Pyrulina gutta* d'Orbigny, 1826 – Hermelin, p. 45, Pl. 5:13.

*Description.* – Test free, fusiform; chambers biserial. Wall calcareous, hyaline. Sutures flush with surface. Aperture radiate.

*Distribution at Sites 502A and 503B.* – *Pyrulina gutta* is represented by a single specimen at Site 502A (80.46 m). It is absent at Site 503B.

## Genus *Pyrulinoides* Marie, 1941

### *Pyrulinoides* sp. 1

Fig. 16A

*Synonymy.* – □1989 *Pyrulinoides* sp. 1 – Hermelin, p. 45, Pl. 5:14

*Description.* – Test free, elongate, acuminate towards both ends, almost circular in transverse section. Chambers biserially arranged, embracing. Wall calcareous, hyaline. Sutures flush with surface. Aperture radiate.

*Distribution at Sites 502A and 503B.* – *Pyrulinoides* sp. 1 is represented by scattered occurrences.

### *Pyrulinoides* sp. 2

Fig. 16B

*Synonymy.* – □1989 *Pyrulinoides* sp. 2 – Hermelin, p. 45, Pl. 5:15.

*Description.* – Test free, elongate, acuminate towards both ends, slightly compressed. Chambers biserially arranged. Sutures flush with surface. Wall calcareous, hyaline. Aperture radiate.

*Distribution at Sites 502A and 503B.* – *Pyrulinoides* sp. 2 is a rare species at both Site 502A and Site 503B.

### *Pyrulinoides* sp. 3

Fig. 16C

*Description.* – Test free, elongate, acuminate towards both ends, almost circular in cross-section. Chambers biserially arranged. Sutures flush with surface. Wall calcareous, hyaline. Aperture radiate.

*Fig. 16.* Scale bar 100 µm. □A. *Pyrulinoides* sp. 1, Side view, Hole 503B, 51.96 m. □B. *Pyrulinoides* sp. 2, Side view, Hole 503B, 68.95 m. □C. *Pyrulinoides* sp. 3, Side view, Hole 502A, 77.69 m. □D. *Saracenaria italica* Defrance, Side view, Hole 503B, 45.80 m. □E. *Saracenaria latifrons* (Brady), Side view, Hole 503B, 48.88 m. □F. *Marginulina*? sp., Side view, Hole 502A, 80.46 m. □G. *Entomorphinoides*? aff. *separata* McCulloch, Side view, Hole 502A, 68.42 m. □H. *Glandulina laevigata* d'Orbigny, Side view, Hole 503B, 72.03 m. □I. *Fissurina* cf. *capillosa* Schwager, Side view, Hole 502A, 71.42 m. □J *Fissurina collifera* Buchner, Side view, Hole 502A,

*Distribution at Sites 502A and 503B.* – *Pyrulinoides* sp. 3 is represented by a single specimen at Site 502A at 77.69 m. It is absent at Site 503B.

## Genus *Saracenaria* Defrance, *in* Blaineville, 1824

### *Saracenaria italica* Defrance, 1824

Fig. 16D

*Synonymy.* – □1824 *Saracenaria italica* n.sp. – Defrance, in Blaineville, p. 176, Pl. 13:6. □1923 *Cristallaria italica* (Defrance) – Cushman, p. 125, Pl. 35:2, 5–7. □1941b *Saracenaria italica* Defrance, 1824 – Leroy (part), p. 76, Pl. 7:21–22 (not Pl. 7:23–24). □1958 *Saracenaria italica* Defrance var. *carapitana* Franklin Defrance, 1824 – Becker & Dusenbury, p. 22, Pl. 2:14a–b. (not *Saracenaria carapitana* Franklin, 1944). □1959 *Saracenaria italica* Defrance, 1824 – Dieci, p. 46, Pl. 4:3. □1960 *Saracenaria italica* Defrance, 1824 – Barker, p. 144, Pl. 68:17, 18, 20–23. □1964 *Saracenaria italica* Defrance, 1824 – Akers & Dorman, p. 52, Pl. 6:20–22. □1976 *Saracenaria italica* Defrance, 1824 – Todd & Low, p. 24, Pl. 3:8a–b. □1985 *Saracenaria italica* Defrance, 1824 – Kohl, p. 49, Pl. 11:3–4.

*Description.* – Test free, large, initial portion closely coiled, planispiral, evolute, periphery subangular, triangular in cross-section, with broad, flat, apertural face. Chambers slightly inflated, rapidly increasing in size as added. Sutures distinct, gently curved. Wall calcareous, hyaline, smooth, finely perforate. Aperture radiate, at peripheral angle.

*Remarks.* – *Saracenaria italica* is larger in size and has sharper lateral edges than *Saracenaria latifrons*.

*Distribution at Sites 502A and 503B.* – *Saracenaria italica* is a rare species with few occurrences at both investigated sites.

### *Saracenaria latifrons* (Brady, 1884)

Fig. 16E

*Synonymy.* – □1884 *Cristellaria latifrons* n.sp. – Brady, p. 544, Pl. 68:19, Pl. 113:11a–b. □1948 *Saracenaria latifrons* (Brady) – Renz, p. 162, Pl. 5:22. □1949 *Saracenaria latifrons* (Brady) – Bermúdez, p. 154, Pl. 8:57–58. □1956 *Saracenaria latifrons* (Brady) – Asano, p. 8, Pl. 3:19. □1964 *Saracenaria latifrons* (Brady) – Leroy, p. 25, Pl. 3:36. □1985 *Saracenaria latifrons* (Brady) – Kohl, 49, Pl. 11:5–6.

*Description.* – Test free, planispiral, early portion coiled and compressed, involute, later uncoiling and flaring, dorsal margin angular, lateral edges rounded, triangular in cross-section, broadest near middle, tapering toward the aperture. Chambers 10–12 in the adult stage, initial chambers closely coiled, later long and narrow. The last four chambers making up two-thirds of the test. Sutures distinct, flush with surface, curved. Wall calcareous, hyaline, smooth, finely perforate. Aperture terminal, radiate, at peripheral angle.

*Distribution at Sites 502A and 503B.* – *Saracenaria latifrons* is represented by a single specimen at Site 503B (33.28 m). It is absent at Site 502A.

## Subfamily Marginulininae Wedekind, 1937

## Genus *Marginulina* d'Orbigny, 1826

### *Marginulina obesa* Cushman, 1923

*Synonymy.* – □1826 *Marginulina glabra* n.sp. – d'Orbigny, p. 259, Modele no. 55. □1913 *Marginulina glabra* d'Orbigny, 1826 – Cushman p. 79, Pl. 23:3. □1923 *Marginulina glabra* d'Orbigny var. *obesa* – Cushman, p. 128, Pl. 37:1. □1950 *Marginulina glabra* d'Orbigny, 1826 – Cushman & McCulloch (part), p. 308, Pl. 40:6 (not Pl. 40:7–8) (not d'Orbigny, 1826). □1960 *Marginulina obesa* Cushman, 1923 – Barker, p. 136, Pl. 65:5–6. □1974 *Marginulina obesa* Cushman, 1923 – LeRoy & Levinson, p. 8, Pl. 4:3–4. □1985 *Marginulina obesa* Cushman, 1923 – Kohl, p. 51, Pl. 12:3. □1990 *Marginulina obesa* Cushman, 1923 – Thomas, p. 227. □1990 *Marginulina obesa* Cushman, 1923 – Ujiié, p. 20, Pl. 6:2a–b; 3.

*Description.* – Test free, elongate, early portion slightly coiled, latter portion rectilinear, rounded in cross-section. Chambers inflated, 5–6 in the adult. Sutures indistinct, slightly depressed on few chambers. Wall calcareous, finely perforate, smooth. Aperture radiate, with a short neck.

*Distribution at Sites 502A and 503B.* – Only one specimen of this species was encountered at Site 502A (46.87 m), whereas it was not observed in the census counts of Site 503B.

## *Marginulina*? sp.

Fig. 16F

*Synonymy.* – □1990 *Marginulina*? sp. – Ujiié, p. 21, Pl. 6:8; 9a–b.

*Description.* – Test free, elongate, biserial, tapering toward each end, circular in transverse section, broadest at the middle of the test. Chambers strongly overlapping and increasing in size. Sutures distinct, flush with surface. Wall calcareous, smooth, finely perforate. Aperture terminal, central, radiate, with short neck.

*Distribution at Sites 502A and 503B.* – *Marginulina*? sp. is a rare species with few occurrences at Site 502A. It is absent at Site 503B.

## *Entomorphinoides*? aff. *separata* McCulloch, 1977

Fig. 16G

*Synonymy.* – □1977 *Entomorphinoides* (?) *separata* – McCulloch, p. 213, Pl. 87:15, 31–35, 37. □1990 *Entomorphinoides* aff. *separata* McCulloch, 1977 – Ujiié, p. 21, Pl. 7:1a–b; 2.

*Description.* – Test free, compressed, elongate, both ends broadly rounded, length about twice as long as broad, almost uniform width, periphery smooth, rounded. Chambers distinct, alternating in a modified biserial organisation with anterior end a shallow uniserial form, posterior end tapering, embracing preceding chambers laterally, ending at center basally. Wall calcareous, thin, hyaline, finely perforate. Aperture fissurine with weak dentition at inner margin of lips, which are flush with contour of anterior end, entosolenian tube short, centered.

*Remarks.* – The specimens in this study more closely resemble the ones described by Ujiié (1990) than those of the typical species.

*Distribution at Sites 502A and 503B.* – *Entomorphinoides*? aff. *separata* is a very rare species with few occurrences at Site 502A. It is absent at Site 503B.

# Family Glandulinidae Reuss, 1860

# Subfamily Glandulininae Reuss, 1860

# Genus *Glandulina* d'Orbigny, 1839

## *Glandulina laevigata* d'Orbigny, 1826

Fig. 16H

*Synonymy.* – □1826 *Nodosaria* (*Glanduline*) *laevigata* sp.nov. – d'Orbigny, p. 252, Pl. 10:1–3. □1923 *Nodosaria* (*Glandulina*) *laevigata* d'Orbigny, var. *torrida* Cushman, p. 65, Pl. 12:10. □1930 *Glandulina laevigata* d'Orbigny – Cushman & Ozawa, p. 143, Pl. 40:1a–b. □1933c *Glandulina laevigata* d'Orbigny – Cushman, p. 41, Pl. 9:14a–b. □1945 *Glandulina laevigata* d'Orbigny – Cushman & Todd, p. 34, Pl. 5:19. □1953 *Glandulina laevigata* d'Orbigny – Loeblich & Tappan, p. 81, Pl. 16:2–5. □1958 *Glandulina laevigata* d'Orbigny – Becker & Dusenbury, p. 25, Pl. 3:23. □1960 *Rectoglandulina torrida* (Cushman) – Barker, p. 128, Pl. 61:20–22. □1964 *Rectoglandulina laevigata* (d'Orbigny) – Leroy, p. 23, Pl. 14:29–30. □1985 *Glandulina laevigata* d'Orbigny – Kohl, p. 54, Pl. 14:2. □1985 *Glandulina laevigata* d'Orbigny – Taylor *et al.*, p. 20, Pl. 1:1–4. □1990 *Pandaglandulina torrida* (Cushman) – Ujiié, p. 21, Pl. 6:4a–b, 5, 6.

*Description.* – Test free, elongate to ovate, tapering toward each end, circular in cross-section. Chambers in early portion biserially arranged and later uniserial, strongly overlapping and rapidly increasing in size as added. Sutures distinct, flush with surface. Wall calcareous, generally opaque except for narrow hyaline band adjacent to aperture, smooth or slightly striated. Aperture terminal, central, radiate, with short, straight entosolenian tube.

*Remarks.* – The major characteristics of this species are the initial biserial stage of the microspheric form and the short, straight entosolenian tube.

*Distribution at Sites 502A and 503B.* – Only one specimen of this species was encountered at Site 503B (72.03 m), whereas it was not observed in the census counts of Site 502A.

## Subfamily Oolininae Loeblich & Tappan, 1961

## Genus *Fissurina* Reuss, 1850

The genus *Fissurina* has typically a compressed test, which is rounded to ovatein outline and may posses a keel. The aperture may be slitlike to oval or rounded with an entosolenian tube projecting inward from the aperture.

## *Fissurina auriculata* (Brady, 1881)

*Synonymy.* – □1881 *Lagena auriculata* n.sp. – Brady, p. 61. □1884 *Lagena auriculata* Brady, 1881 – Brady (part), p. 487, Pl. 60:29 (not 31, 33). □1960 *Fissurina auriculata* (Brady) – Barker, p. 126, Pl. 60:29. □1982 *Fissurina auriculata auriculata* (Brady) – Boltovskoy & de Kahn, p. 423, Pl. 1:23–26. □1989 *Fissurina auriculata* (Brady) – Hermelin, p. 46, Pl. 6:2–3.

*Description.* – Test free, unilocular, pyriform, slightly compressed, with an elliptic tube at the periphery on each side of the base. Wall calcareous, smooth, hyaline, finely perforate. Aperture terminal, rounded.

*Distribution at Sites 502A and 503B.* – *Fissurina auriculata* is a rare species with few occurrences at both Site 502A and Site 503B.

## *Fissurina* cf. *capillosa* Schwager

Fig. 16I

*Synonymy.* – □1986 *Fissurina capillosa* (Schwager) – Boersma, p. 1019, Pl. 4:6–7. □1989 *Fissurina* cf. *F. capillosa* Schwager – Hermelin, p. 46, Pl. 6:4–5.

*Description.* – Test free, unilocular, pyriform, compressed, ovate central body, very wide peripheral keel. Wall calcareous, hyaline, inner part of keel and base of the neck ornamented with a reticulate pattern. Aperture terminal, rounded.

*Remarks.* – These specimens of *Fissurina capillosa* are very similar to the specimen figured by Hermelin (1989) in having ornamentation along the entire margin, but not along the central body as in the type specimen. I suggest that this difference represents phenotypic variation within *Fissurina capillosa*.

*Distribution at Sites 502A and 503B.* – *Fissurina* cf. *capillosa* is a rare species with scattered occurrences at both Site 502A and Site 503B. It is more abundant at Site 502A than at Site 503B.

## *Fissurina castrensis* (Schwager, 1866)

*Synonymy.* – □1866 *Lagena castrensis* – Schwager, p. 208, Pl. 5:22. □1989 *Fissurina castrensis* (Schwager) – Hermelin, p. 46, Pl. 6:6–7.

*Description.* – Test free, unilocular, circular at the central part of the test, compressed, three parallel keels encircle the test, median keel with short spines in the basal part. Wall calcareous, hyaline, finely perforate. Aperture terminal, round, on a short ornamented neck.

*Remarks.* – The specimens of *F. castrensis* encountered in this investigation at Sites 502A and 503B, resemble in most, aspects that figured by Hermelin (1989) but are not ornamented with spines in the central part of the test. This probably represents phenotypic variation without taxonomic significance.

*Distribution at Sites 502A and 503B.* – *Fissurina castrensis* is a rare species with scattered occurrences in both Site 502A and Site 503B.

## *Fissurina clathrata* (Brady, 1884)

*Synonymy.* – □1884 *Lagena clathrata* n.sp. – Brady, p. 484, Pl. 60:4. □1901 *Lagena orbignyana* (Seguenza) var. *clathrata* Brady – Millett, p. 628, Pl. 14:23. □1913 *Lagena orbignyana* (Seguenza) var. *clathrata* Brady – Cushman, p. 44, Pl. 11:4. □1933c *Lagena orbignyana* (Seguenza) var. *clathrata* Brady – Cushman, p. 28, Pl. 7:6, 7?. □1950 *Fissurina clathrata* (Brady) – Parr, p. 310. □1960 *Fissurina clathrata* (Brady)– Barker, p. 124, Pl. 60:4.

*Description.* – Test free, unilocular, with a central body and a broad peripheral keel, compressed, ovate in cross-section, body circular in side view, keel thin at base, thickening toward the apertural end and forming a bulbous structure. Wall calcareous, hyaline, smooth, finely perforate, central body ornamented with longitudinal costae.

*Remarks.* – *Fissurina clathrata* has been regarded as a variety of *Fissurina orbignyana* (Seguenza).

*Distribution at Sites 502A and 503B.* – *Fissurina clathrata* is a very rare species with only a few occurrences at Site 502A. It is absent at Site 503B.

## *Fissurina collifera* Buchner, 1940

Fig. 16J

*Synonymy.* – □1940 *Fissurina pseudoorbignyana* Buchner, forma *collifera* – Buchner, pp. 460–461, Pl. 10:159–160. □1990 *Fissurina* cf. *collifera* Buchner – Ujiié, pp. 25–26, Pl. 9:2a–b.

*Description.* – Test free, unilocular, with a central body and a broad peripheral keel, ovate in cross-section, body circular in side view, keel thin, the apertural end and forming a bulbous structure. Wall calcareous, hyaline, smooth, finely perforate.

*Distribution at Sites 502A and 503B.* – *Fissurina collifera* is represented by a single specimen at Site 502A at 111.48 m. It is absent at Site 503B.

## Fissurina crebra (Matthes, 1960)

*Synonymy.* – □1960 *Fissurina crebra* (Matthes) – Barker, p. 122, Pl. 59:6a–b. □1982 *Fissurina crebra* (Matthes) – Boltovskoy & de Kahn, p. 424, Pl. 2:17–20. □1989 *Fissurina crebra* (Matthes) – Hermelin, p. 47, Pl. 6:11–12. □1990 *Fissurina crebra* (Matthes) – Ujiié, p. 26, Pl. 9:5a–b.

*Description.* – Test free, unilocular, compressed, ovate in side view, periphery with a narrow keel, pointed base. Wall calcareous, hyaline, smooth, finely perforate. Aperture terminal, elliptical, at end of short neck.

*Distribution at Sites 502A and 503B.* – *Fissurina crebra* is absent at Site 502A and is a very rare species with only a few occurrences at Site 503B (above 64.85 m).

## *Fissurina fimbriata* (Brady, 1881)

Fig. 17A–B, F

*Synonymy.* – □1881 *Lagena fimbriata* n.sp. – Brady, p. 61. □1884 *Lagena fimbriata* Brady, 1881 – Brady, pp. 486–487, Pl. 60:26–28. □1913 *Lagena fimbriata* Brady, 1881 – Cushman, p. 30, Pl. 14:8. □1931 *Lagena (Entosolenia) fimbriata* Brady – Wiesner, p. 122, Pl. 19:232. □1950 *Fissurina fimbriata* (Brady) – Parr, p. 307. □1960 *Fissurina fimbriata* (Brady)– Barker, p. 126, Pl. 60:16–28. □1982 *Fissurina fimbriata* (Brady) *F. typica* Boltovskoy & de Kahn, p. 425, Pl. 3:13–14. □1989 *Fissurina fimbriata* (Brady) – Hermelin, p. 47, Pl. 6:15–16.

*Description.* – Test free, unilocular, pyriform, rounded, broad at the base, slightly compressed, a thin vertical structure, often with a very fine longitudinal striation, encircles the oval base. Aperture terminal, rounded.

*Distribution at Sites 502A and 503B.* – *Fissurina fimbriata* is a rare species at Site 502A, occurring in many samples but in low abundances mainly above 112.42 m. It is represented by only one specimen at Site 503B (63.27 m).

## *Fissurina kerguelenensis* Parr, 1950

Fig. 17C

*Synonymy.* – □1950 *Fissurina kerguelenensis* n.sp. – Parr, p. 305, Pl. 8:7a–b. □1960 *Fissurina kerguelenensis* Parr, 1950 – Barker, p. 122, Pl. 59:8–11. □1982 *Fissurina staphyllearia staphyllearia* (Schwager) – Boltovskoy & de Kahn, p. 431, Pl. 7:13–14 (not *Fissurina staphyllearia* Schwager, 1866). □1989 *Fissurina kerguelenensis* Parr, 1950 – Hermelin, p. 48, Pl. 6:17–18.

*Description.* – Test free, unilocular, compressed, rounded in side view, slightly produced apertural end, a keel extends around the test, 3–10 spines at the basal part of the test. Wall calcareous, hyaline, smooth, finely perforate, transparent. Aperture a narrow slit in the plane of compression.

*Distribution at Sites 502A and 503B.* – *Fissurina kerguelenensis* is a very rare species with few occurrences at both investigated sites.

## *Fissurina marginata* (Montagu, 1803)

Fig. 17D–E

*Synonymy.* – □1784 *Serpula (Lagena) marginata* – Walker & Boys, p. 2, Pl. 1:7 [nomen nudum]. □1803 *Vermiculum marginatum* – Montagu, p. 524. (refer to Walker & Boys, 1784, pl. 1:7). □1884 *Lagena marginata* (Walker & Boys) [*sic*] – Brady (part), pp. 476–477, Pl. 59:21–22 (not 23). □1941 *Fissurina marginata* (Walker & Boys) [*sic*].– Galloway & Heminway, p. 353, Pl. 11:4a–b. □1953 *Fissurina marginata* (Montagu) – Loeblich & Tappan, p. 77, Pl. 14:6-9. □1964 *Fissurina marginata* (Walker & Boys) [*sic*] – Feyling-Hanssen, p. 315, Pl. 15:22. □1985 *Fissurina marginata* (Montagu) – Hermelin & Scott, p. 208, Pl. 2:17. □1985 *Fissurina marginata* (Montagu) – Kohl, pp. 55–56, Pl. 15:5a–b. □1989 *Fissurina marginata* (Montagu) – Hermelin, p. 48, Pl. 7:1–2.

*Description.* – Test free, unilocular, compressed, round to ovate in side view with a slightly produced apertural neck, periphery with a thin keel. Entosolenian tube running from aperture along wall of central body down to basal part of test. Wall calcareous, hyaline, smooth, finely perforate. Aperture terminal, on short neck, rounded.

*Remarks.* – According to Hermelin (1989) *Fissurina marginata* has incorrectly been accredited to Walker & Boys (1784) by many authors. Since Walker & Boys (1784) did not follow the convention of binominal nomenclature, their specific names should be rejected (ICZN, 1959).

*Distribution at Sites 502A and 503B.* – *Fissurina marginata* occurs in many samples throughout the studied interval of both sites, but in relatively low abundances.

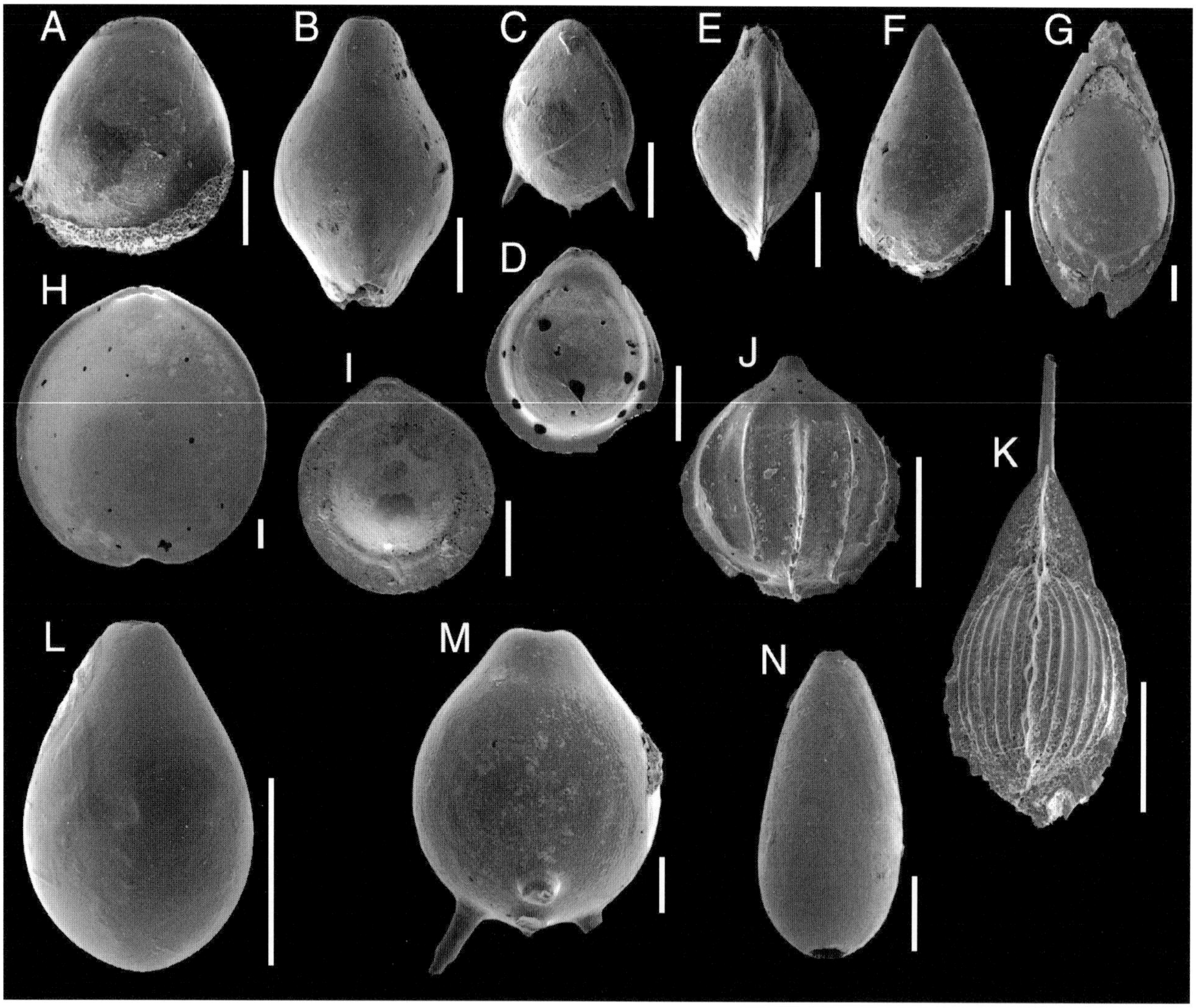

*Fig. 17.* Scale bar 100 µm. □A, B, F. *Fissurina fimbriata* (Brady); A, Side view, Hole 502A, 64.51 m; B, Edge view, Hole 503B, 63.27 m; C, Side view, Hole 502A, 84.96 m. □C. *Fissurina kerguelenensis* Parr, Side view, Hole 502A, 83.99 m. □D, E. *Fissurina marginata* (Montagu); E, Side view, Hole 502A, 90.94 m; F, Edge view, Hole 502A, 98.81 m. □G. *Fissurina semimarginata* (Reuss), Side view, Hole 502A, 49.67 m. □H, I. *Fissurina wiesneri* Barker; H, Side view, Hole 502A, 64.51 m; I, Side view, Hole 502A, 91.85 m. □J. *Oolina alifera* (Reuss), Side view, Hole 502A, 66.51 m. □K. *Oolina desmophora* (Jones), Side view, Hole 503B, 65.37 m. □L. *Oolina globosa* (Montagu), Side view, Hole 502A, 125.35 m. □M. *Oolina setosa* (Earland), Side view, Hole 502A, 66.51 m. □N. *Oolina* sp. 1, Side view, Hole 502A, 108.66 m.

## *Fissurina semimarginata* (Reuss, 1870)

Fig. 17G

*Synonymy.* – □1870 *Lagena* sp. (Nos. 64–65) – von Schlict, p. 11, Pl. 4:4–6, 10–12. □1870 *Lagena marginata* Williamson var. *semimarginata* – Reuss, p. 468. □1884 *Lagena marginata* Williamson var. *semimarginata* Reuss, 1870 – Brady, p. 446, Pl. 59:17, 19. □1931 *Lagena* (*Entosolenia*) *marginata* var. *semimarginata* Reuss – Wiesner, p. 120, Pl. 19:224. □1953 *Fissurina semimarginata* (Reuss) –

Loeblich & Tappan, p. 78, Pl. 14:3. □1960 *Fissurina semimarginata* (Reuss) – Barker, p. 122, Pl. 59:17–18.

*Description.* – Test free, unilocular, compressed, ovate in outline, with rounded base. Wall calcareous, finely and distinctly perforate, surface smooth, with a slight marginal keel in the upper part. Aperture elongate, produced on a clear necklike extension of the chamber, and with a long entosolenian tube attached to the wall of the central body.

*Distribution at Sites 502A and 503B. – Fissurina semima-rginata* occurs in many samples at Site 502A but in relatively low abundances. It is rare with few occurrences at Site 503B.

## Fissurina wiesneri Barker, 1960

Fig. 17H–I

*Synonymy. –* □1960 *Fissurina wiesneri* nom.nov. – Barker, p. 124, Pl. 59:23. □1989 *Fissurina wiesneri* Barker, 1960 – Hermelin, pp. 49–50, Pl. 7:10. □1990 *Fissurina wiesneri* Barker, 1960 – Ujiié, p. 26, Pl. 9:3a–b; 4a–b.

*Description. –* Test free, unilocular, central body round, compressed, a very wide keel encirles the test. Wall calcareous, hyaline, smooth, finely perforate. Aperture terminal, rounded, on a neck as long as the keel is wide.

*Remarks. –* Characteristic for *Fissurina wiesneri* is the wide keel, which surrounds the test.

*Distribution at Sites 502A and 503B. – Fissurina wiesneri* is a rare species with a few occurrences at both Site 502A and Site 503B.

## Genus *Oolina* d'Orbigny, 1839

## Oolina alifera (Reuss, 1870)

Fig. 17J

*Synonymy. –* □1870 *Lagena alifera* nom.nov. – Reuss, p. 467, Text-figs. 15–16, 21–22. □1985 *Oolina* cf. *alifera* (Reuss) – Kohl, p. 57, Pl. 16:2. □1989 *Oolina alifera* (Reuss) – Hermelin, p. 55, Pl. 10:1.

*Description. –* Test free, unilocular, rounded in cross-section, elliptical in side view, base slightly truncated. Wall calcareous, hyaline, finely perforate, ornamented with five to six prominent longitudinal costae, originating at a ring at the base and merging into a smooth collar. Aperture terminal, round opening at the end of a short neck that projects from the collar, with a short entosolenian tube projecting into the chamber.

*Remarks. –* The original *Oolina alifera* differs from the recently described *Oolina* cf. *alifera* (Kohl, 1985) in having fewer costae, 8–9 rather than 12–15. Because of this difference, some of the specimens found in this study may be classified as *Oolina* cf. *alifera,* but no separate taxonomic status is given to this form here.

*Distribution at Sites 502A and 503B. – Oolina alifera* is a rare species with few occurrences above 66.51 m at Site 502A. It is absent at Site 503B.

## Oolina desmophora (Jones, 1874)

Fig. 17K

*Synonymy. –* □1874 *Lagena vulgaris* Williamson var. *desmophora* – Jones, p. 54, Pl. 19:23–24, □1884 *Lagena desmophora* Jones – Brady, pp. 468–469, Pl. 58:42–43. □1921 *Lagena desmophora* Jones – Cushman, p. 27, Pls. 12:5a–b; 13:3a–b. □1960 *Oolina desmophora* (Jones) – Barker, p. 120, Pl. 58:42–43. □1982 *Lagena desmophora* Jones – Boltovskoy & de Kahn, p. 434, Pl. 9:1–3. □1989 *Oolina desmophora* (Jones) – Hermelin, p. 55, Pl. 10:2–3. □1990 *Oolina desmophora* (Jones) – Ujiié, pp. 21–22, Pl. 5:10.

*Description. –* Test free, unilocular, pyriform, rounded in cross-section. Wall calcareous, hyaline, ornamented with five to seven longitudinal costae with depressions at regular intervals in the lower one-half of the test; in between is a secondary costa, which often branches into two. Aperture terminal, round on a long neck, with an entosolenian tube.

*Distribution at Sites 502A and 503B. – Oolina desmophora* is absent at Site 502A and is represented by a single specimen at Site 503B (65.37 m).

## Oolina globosa (Montagu, 1803)

Fig. 17L

*Synonymy. –* □1803 *Vermiculum globosum* – Montagu, p. 503. □1950 *Oolina globosa* (Montagu) – Parr, p. 302. □1960 *Oolina globosa* (Montagu) – Barker, p. 114, Pl. 56:1–3. □1981 *Oolina globosa* (Montagu) – Cole, pp. 75–76, Pl. 19:8. □1982 *Oolina globosa* (Montagu) f. *typica* – Boltovskoy & de Kahn, p. 442, Pl. 12:16–18. □1989 *Oolina globosa* (Montagu)– Hermelin, p. 56, Pl. 10:4. □1990 *Oolina globosa* (Montagu)– Ujiié, p. 22, Pl. 7:6.

*Description. –* Test free, unilocular, circular in cross-section, oval in side view. Wall calcareous, hyaline, smooth. Aperture terminal, with an entosolenian tube extending into the chamber.

*Distribution at Sites 502A and 503B. – Oolina globosa* is a rare species with few occurrences at both investigated sites.

## Oolina setosa (Earland, 1934)

Fig. 17M

*Synonymy. –* □1934 *Lagena globosa* (Montagu) var. *setosa* – Earland, p. 150, Pl. 6:52. □1960 *Oolina globosa* (Montagu) var. *setosa* (Earland) – Barker, p. 116, Pl. 56:33–35. □1964 *Oolina globosa* (Montagu) var. *setosa* [sic] (Earland) – Leroy, p. 26, Pl. 13:44. □1982 *Oolina globosa*

(Montagu) var. *setosa* (Earland) – Boltovskoy & de Kahn, p. 442, Pl. 12:19. □1985 *Oolina setosa* (Earland) – Kohl, p. 58, Pl. 16:6. □1989 *Oolina setosa* (Earland) – Hermelin, p. 57, Pl. 10:6.

*Description.* – Test free, unilocular, subcircular to ovate in side view, circular in cross-section. Wall calcareous, hyaline, body smooth, finely perforate, base ornamented with irregularly arranged short spines. Aperture terminal with an elliptical opening.

*Distribution at Sites 502A and 503B.* – *Oolina setosa* is a rare species with few occurrences above 72.39 m at Site 502A and below 73.82 m at Site 503B.

## *Oolina* sp. 1

Fig. 17N

*Description.* – Test free, unilocular, elongate, circular in cross-section, oval in side view, widest below the middle of the test. Wall calcareous, hyaline, smooth. Aperture terminal, with an entosolenian tube extending into the chamber.

*Distribution at Sites 502A and 503B.* – *Oolina* sp. 1 is a rare species with few occurrences at both investigated sites.

## Genus *Parafissurina* Parr, 1947

The genus *Parafissurina* has a test with a single ovate chamber, which is often compressed, with a subterminal arched aperture at one side of the test with an overhanging hoodlike extension of the test wall.

## *Parafissurina sublata* Parr, 1950

Fig. 18A

*Synonymy.* – □1950 *Parafissurina sublata* n.sp. – Parr, p. 319, Pl. 10:11a–c. □1982 *Parafissurina sublata* Parr, 1950 – Boltovskoy & de Kahn, p. 446, Pl. 15:5. □1989 *Parafissurina sublata* Parr, 1950 – Hermelin, p. 57, Pl. 10:8.

*Description.* – Test free, unilocular, circular, compressed in cross-section, with a narrow peripheral keel. Wall calcareous, hyaline, smooth. Aperture a hooded slit at one side of the test that extends to the base of the test along the center of the wall.

*Distribution at Sites 502A and 503B.* – Only one specimen of this species was encountered at Site 502A (133.26 m). It is rare with few occurrences at Site 503B.

## *Parafissurina tectulostoma* Loeblich & Tappan, 1953

*Synonymy.* – □1953 *Parafissurina tectulostoma* n.sp. – Loeblich & Tappan, p. 81, Pl. 14:17a–c. □1981 *Parafissurina tectulostoma* Loeblich & Tappan, 1953 – Cole, p. 85, Pl. 19:36. □1982 *Parafissurina tectulostoma* Loeblich & Tappan, 1953 – Boltovskoy & de Kahn, p. 446, Pl. 15:8–9. □1989 *Parafissurina tectulostoma* Loeblich & Tappan, 1953 – Hermelin, p. 57, Pl. 10:9. □1990 *Parafissurina tectulostoma* Loeblich & Tappan, 1953 – Thomas *et al.*, p. 227, Pl. 5:9.

*Description.* – Test free, unilocular, elongate, rounded in cross-section, widest below the middle part of the test. Wall calcareous, hyaline, smooth. Aperture a hooded subterminal slit with an entosolenian tube extending down to about one-half the length of the test.

*Distribution at Sites 502A and 503B.* – *Parafissurina tectulostoma* is represented by a single specimen at Site 502A (47.76 m) and is absent at Site 503B.

## *Parafissurina tricarinata* Parr, 1950

*Synonymy.* – □1950 *Parafissurina tricarinata* n.sp. – Parr, pp. 319–320, Pl. 10:16–18. □1989 *Parafissurina tricarinata* Parr, 1950 – Hermelin, pp. 57–58, Pl. 10:10.

*Description.* – Test free, unilocular, compressed, ovate in cross-section, circular in side view. Wall calcareous, hyaline, smooth, finely perforate, periphery with distinct keel, and two additional keels on the lower part of the test, one on either side. Aperture terminal, slit-like, slightly curved on one side and with a raised lip on either edge, lip on one side slightly higher than the other, with short entosolenian tube.

*Distribution at Sites 502A and 503B.* – Only one specimen of this species was represented at Site 503B (56.39 m). It was absent at Site 502A.

## *Parafissurina uncifera* (Buchner, 1940)

*Synonymy.* – □1940 *Lagena uncifera* n.sp. – Buchner, p. 531, Pl. 26:554–555. □1982 *Parafissurina uncifera* (Buchner) – Boltovskoy & de Kahn, p. 446, Pl. 15:10–11. □1982 *Parafissurina uncifera* (Buchner) – Hermelin, 1989, p. 58, Pl. 10:12–13.

*Description.* – Test free, unilocular, globular. Wall calcareous, hyaline, smooth, thick. Aperture rounded, with a raised rim which is extended on one side.

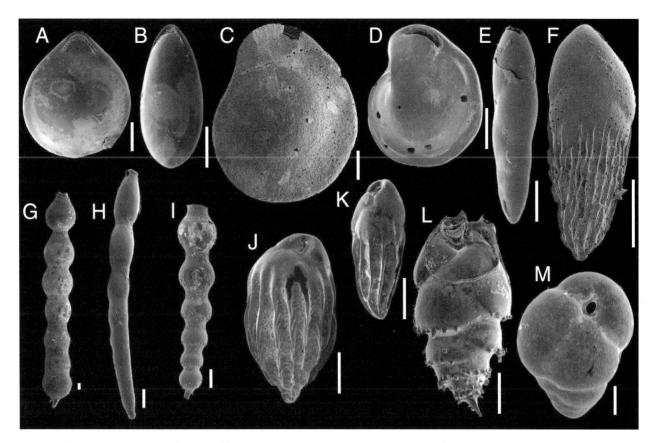

*Fig. 18.* Scale bar 100 µm. □A. *Parafissurina sublata* Parr, Side view, Hole 503B, 73.82 m. □B. *Parafissurina* sp. 1 Hermelin, Side view, Hole 503B, 91.37 m. □C, D. *Hoeglundina elegans* (d'Orbigny); C, Spiral view, Hole 502A, 48.80 m; D, Umbilical view, Hole 502A, 113.42 m. □E. *Bolivina seminuda* Cushman, Side view, Hole 502A, 108.66 m. □F *Bolivina subaenariensis* (Cushman), Side view, Hole 502A, 135.33 m. □G. *Siphonodosaria abyssorum* (Brady), Side view, Hole 503B, 76.36 m. □H. *Siphonodosaria consobrina* (d'Orbigny), Side view, Hole 503B, 67.26 m. □I. *Siphonodosaria lepidula* (Schwager), Side view, Hole 503B, 64.85 m. □J, K. *Bulimina alazanensis* Cushman; J, Side view, Hole 502A, 78.45 m; K, Side view, Hole 502A, 78.45 m. □L. *Bulimina marginata* (d'Orbigny), Side view, Hole 503B, 87.44 m. □M. *Bulimina translucens* Parker, Apertural view, Hole 503B, 48.88 m.

*Distribution at Sites 502A and 503B.* – *Parafissurina uncifera* is represented by one single specimen at Site 502A (77.69 m). It is rare with few occurrences at Site 503B.

## *Parafissurina* sp. 1

Fig. 18B

*Synonymy.* – □1989 *Parafissurina* sp. 1 – Hermelin, p. 58, Pl. 10:11.

*Description.* – Test free, unilocular, elongate, compressed in transverse-section. The test is widest below the middle part, and has a rounded base. Wall calcareous, hyaline, smooth. Aperture hooded, with an entosolenian tube extending down along the rear wall to about one-half the length of the test.

*Remarks.* – *Parafissurina* sp. 1 differs from *P. tectulostoma* in being compressed in cross-section rather than round (Hermelin 1989).

*Distribution at Sites 502A and 503B.* – *Parafissurina* sp. 1 is a rare species with few occurrences.

## Superfamily Robertinacea Reuss, 1850

## Family Epistominidae Wedekind, 1937

## Genus *Hoeglundina* Brotzen, 1948

## *Hoeglundina elegans* (d'Orbigny, 1826)

Fig. 18C–D

*Synonymy.* – □1826 *Rotalia* (*Turbinulina*) *elegans* – d'Orbigny, p. 276, Modeles no. 54 (figured in Parker, Jones & Brady, 1871, Pl. 12:142). □1930 *Epistomina elegans* (d'Orbigny) Cushman & Jarvis, p. 365, Pl. 34:1a–c. □1941a *Epistomina elegans* (d'Orbigny) – Leroy, p. 38, Pl. 6:5–7. □1951 *Epistomina elegans* (d'Orbigny) – Asano, p.

17:130–131. □1951 *Höglundina elegans* (d'Orbigny) – Phleger & Parker, p. 22, Pl. 12:1a–b. □1964 *Höglundina elegans* (d'Orbigny)– Leroy, p. 38, Pl. 6:27–28. □1966 *Höglundina elegans* (d'Orbigny)– Belford, p. 190, Pl. 36:. □1985 *Höglundina elegans* (d'Orbigny)– Kohl, p. 59, Pl. 14:4–5. □1990 *Höglundina elegans* (d'Orbigny)– Thomas *et al.*, p. 227, Pl. 7:12.

*Description.* – Test free, trochospiral, involute, biconvex in cross-section; periphery with a keel, circular in side view. Chambers distinct, 7–9 in the last whorl, increasing gradually in size as added. Sutures distinct, flush with surface, strongly oblique on the spiral side, obliquely radial on the umbilical side. Wall calcareous, hyaline, finely perforate, smooth. Aperture an elongate slit, extending the breadth of ultimate chamber and parallel to peripheral margin on umbilical side.

*Remarks.* – *Hoeglundina elegans* (the type species for *Hoeglundina* Brotzen 1948) has a characterisitic distinctive supplementary aperture, which appears as an arched, linear slit, extending the breadth of the final chamber, and a smooth, large, keeled, thickly calcified, biconvex and essentially imperforate test.

*Ecology.* – *Hoeglundina elegans,* the only widely distribuuted aragonitic deep-sea foraminifer, is widespread in the Atlantic at depth ranging from 42 to 4,330 m (Phleger & Parker 1951). In the western South Atlantic Ocean, it is found at water depth between 2,500 and 3,500 m (Lohmann 1978). In the Gulf of Mexico, its upper depth limit is 65 m (Parker 1954), in Florida Bay (Bock 1971) and on the North Carolina shelf (Schnitker 1971). Pflum & Frerichs (1976) found *H. elegans* increase in size with increased water depth; from about 1.0 mm in the middle bathyal zone to about 2.0 mm in the lower bathyal and abyssal zones in the Gulf of Mexico. *Hoeglundina elegans* has previously been regarded as a marker for NADW, where the salinity ranges between 34.5 and 35.8‰, and the temperature from about 2 to 3°C (e.g., Lohmann 1978; Schnitker 1979; Murray 1991). Miller & Lohmann (1982) found this species more abundant in areas with high organic content in the sediment. On the Ontong Java Plateau, *H. elegans* is reported to have its maxima in the transitional between glacials and inglacials (Burke *et al.* 1993). The high abundance of *Hoeglundina* in the transitional zone may reflect periods of increased carbonate preservation (Burke *et al.* 1993); this species has an aragonatic shell (Bandy 1954). According to Burke (1981), and Hermelin & Shimmield (1990) *H. elegans* is related to low oxygen conditions and is also considered to be an epifaunal species that lives in the 0–1 cm interval (Corliss, 1985, 1991).

*Distribution at Sites 502A and 503B.* – *Hoeglundina elegans* is represented by only a few specimens at Site 502A. It is absent at Site 503B.

# Superfamily Buliminacea Jones, 1875

# Family Bolivinitidae Cushman, 1927

# Genus *Bolivina* d'Orbigny, 1839

## *Bolivina catanensis* Seguenza, 1862

*Synonymy.* – □1862b *Bolivina catanensis* n.sp. – Seguenza, p. 29, Pl. 2:3. □1958 *Bolivina catanensis* Seguenza, 1862b – Parker, p. 260, Pl. 2:10–11. □1964 *Bolivina catanensis* Seguenza, 1862b – Akers & Dorman, p. 24, Pl. 8:3, 21. □1978 *Bolivina catanensis* Seguenza, 1862b – Wright, p. 710, Pl. 1:16–17.

*Description.* – Test free, compressed, biserial, twice as long as broad, widest near the apertural end. Chambers six to nine pairs in the adult; chambers broader than high, increasing gradually in size. Sutures limbate, raised, slightly curved. Wall calcareous, hyaline, finely perforate. Aperture terminal, elongate, oblique, extending from the base of the ultimate chamber to a terminal position.

*Remarks.* – *Bolivina catanensis* differs from *Bolivina goesii* in being more elongated and less compressed.

*Distribution at Sites 502A and 503B.* – *Bolivina catanensis* is represented by a single specimen at Site 502A (120.27 m) and is absent at Site 503B.

## *Bolivina seminuda* Cushman, 1911

Fig. 18E

*Synonymy.* – □1911 *Bolivina seminuda* n.sp. – Cushman, p. 345, Text-fig. 55. □1931 *Bolivina seminuda* Cushman, 1911 – Hada, p. 132, Text-fig. 89. □1937c *Bolivina seminuda* Cushman, 1911 – Cushman, p. 142, Pl. 18:13–15. □1942 *Bolivina seminuda* Cushman, 1911 – Cushman, p. 26, Pl. 7:6. □1942 *Bolivina seminuda* Cushman, 1911 – Cushman & McCulloch, p. 210, Pl. 25:14. □1947 *Bolivina pseudopunctata* n.sp. – Höglund, pp. 273–274, Pls. 24:5; 32:23–24, Text-figs. 280–281, 287. □1953 *Bolivina pseudopunctata* Höglund, 1947 – Phleger *et al.*, p. 36, Pl. 7:20–21. □1963 *Bolivina seminuda* Cushman, 1911– Smith, pp. 15–16, Pl. 29:1–7. □1980 *Bolivina seminuda* Cushman, 1911 – Ingle *et al.*, p. 131, Pl. 1:5. □1981 *Bolivina seminuda* Cushman, 1911 – Resig, Pl. 5:14. □1985 *Bolivina pseudopunctata* Höglund, 1947– Hermelin & Scott, pp. 203–204, Pl. 2:18. □1986 *Bolivina seminuda* Cushman, 1911 – Kurihara & Kennett, Pl. 2:9–10. □1989 *Bolivina seminuda* Cushman, 1911 – Hermelin, p. 60, Pl. 10:17–18. □1990 *Bolivina pseudopunctata* Höglund, 1947 – Schröder-Adams *et al.*, p. 24, Pl. 6:8.

*Description.* – Test free, biserial, elongate, slightly compressed. Chambers distinct. Sutures distinct, depressed. Wall calcareous, hyaline, smooth in the upper part, whereas lower two-thirds are coarsely perforate. Aperture oval, extending from base of ultimate chamber to a terminal position.

*Remarks.* – Hermelin (1989) suggested that *Bolivina pseudopunctata*, described by Höglund (1947) as an Atlantic species, appears to be in most respects a junior synonym of *B. seminuda*. According to Hermelin (1989), *B. seminuda* and *B. pseudopunctata* appear to be identical, except for their reported geographical distribution; *B. seminuda* is reported mostly from the Pacific, whereas *B. pseudopunctata* is reported mostly from the Atlantic. This supports a synonymization of the two forms.

*Ecology.* – Smith (1963, 1964) reported *Bolivina seminuda* from the middle and lower bathyal zones off the west coast of Central America. In the Gulf of California this species is reported in the neritic through the lower bathyal zones (Bandy, 1961). *Bolivina* has frequently been associated with low-oxygen conditions in the bottom waters (Rathburn & Corliss 1994). It is also considered to be shallow-infaunal with maxima between 1.5 and 4 cm (Corliss, 1985, 1991; Corliss & Chen 1988).

*Distribution at Sites 502A and 503B.* – *Bolivina seminuda* is represented by a few specimens at Site 502A (108.66 m) and by a single specimen at Site 503B (59.00 m).

## *Bolivina subaenariensis* (Cushman, 1922)

Fig. 18F

*Synonymy.* – □1922 *Bolivina subaenariensis* sp.nov. – Cushman, p. 46, Pl. 7:6. □1937c *Bolivina subaenariensis* Cushman, 1922 – Cushman, p. 155, Pl. 18:26–28. □1957 *Bolivina subaenariensis* Cushman, 1922 – Todd & Brönnimann, p. 34, Pl. 8:19–20. □1964 *Bolivina subaenariensis* Cushman, 1922 – Akers & Dorman, p. 26, Pl. 8:20, 28. □1971 *Brizalina subaenariensis* (Cushman) – Murray, p. 111, Pl. 45:5–7. □1985 *Brizalina subaenariensis* (Cushman) – Kohl, p. 63, Pl. 17:13.

*Description.* – Test free, biserial, compressed, widest near apertural end, twice as long as broad; periphery acute, carinate, seven to nine pairs of chambers in the adult, increasing gradually in size, curved. Sutures strongly curved, distinct, slightly depressed. Wall calcareous, hyaline, finely perforate, ornamented with two or more costae, the center two extending from the initial end two-thirds the length of test; additional costae shorter, on either side of the other two. Aperture elongate, narrow, surrounded by a rim, with narrow toothplate.

*Distribution at Sites 502A and 503B.* – *Bolivina subaenariensis* is a very rare species with only a few occurrences at Site 502A. It is represented by a single specimen at Site 503B (46.88 m).

# Family Eouvigerinidae Cushman, 1927

# Genus *Siphonodosaria* Silvestri, 1924

In this study, I have followed Loeblich & Tappan (1964) in grouping uniserial eouvigerinids into different genera depending on the shape of the phialine lip and the apertural tooth. *Siphonodosaria* Silvestri (1924) includes forms that have a neck bordered by a dentate phialine lip and a distinct bifid tooth, while *Stilostomella* Guppy (1894) has only a single tooth or slight indentation of the phialine lip.

## *Siphonodosaria abyssorum* (Brady, 1881)

Fig. 18G

*Synonymy.* – □1881 *Nodosaria abyssorum* n.sp. – Brady, p. 63. □1884 *Nodosaria*(?) *abyssorum* Brady – Brady, p. 504, Pl. 63:8–9. □1927a *Siphonodosaria abyssorum* (Brady) – Cushman, p. 69, Pl. 14:20. □1931 *Sagrinnodosaria abyssorum* (Brady) – Jedlitschka, p. 125, Pl. 126:24–25. □1934 *Ellipsonodosaria nuttalli* sp.nov. – Cushman & Jarvis, p. 72, Pl. 10:6a–b. □1952 *Siphonodosaria abyssorum* (Brady) – Stainforth, p. 11, Pl. 1:a–b. □1973 *Siphonodosaria abyssorum* (Brady) – Douglas, Pl. 5:11. □1978 *Stilostomella abyssorum* (Brady) – Boltovskoy, Pl. 7:16. □1989 *Siphonodosaria abyssorum* (Brady) – Hermelin, p. 61, Pl. 11:2–5.

*Description.* – Test free, uniserial, straight, large. Chambers subglobose; proloculus is often the largest chamber. Wall calcareous, smooth, thick, several large, stout spines on basal part of proloculus. Aperture terminal, on a short neck, with a wide phialine lip and a bifid tooth.

*Remarks.* – I agree with Hermelin (1989) that *Ellipsonodosaria nuttalli* (Cushman & Jarvis, 1934) should be regarded as a junior synonym of *Siphonodosaria abyssorum* (Brady).

*Distribution at Sites 502A and 503B.* – *Siphonodosaria abyssorum* is represented by a single specimen at Site 502A (133.26 m). It exhibits more scattered occurrences at Site 503B.

## *Siphonodosaria consobrina* (d'Orbigny, 1846)

Fig. 18H

*Synonymy.* – □1846 *Dentalina consobrina* n.sp. – d'Orbigny, p. 46, Pl. 2:1–3. □1884 *Nodosaria (D.) consobrina* (d'Orbigny) – Brady, p. 501, Pl. 62:23–24. □1951 *Nodogenerina consobrina* (d'Orbigny) – Marks, p. 55. □1952 *Siphonodosaria consobrina* (d'Orbigny) – Stainforth, p. 12. □1960 *Stilostomella consobrina* (d'Orbigny) – Barker, p. 130, Pl. 62:23–24. □1986 *Stilostomella consobrina* (d'Orbigny) – Boersma, Pl. 13:4–5.□1989 *Siphonodosaria consobrina* (d'Orbigny) – Hermelin, p. 61, Pl. 11:6–7.

*Description.* – Test free, uniserial, slender, gradually tapering to the initial end. Chambers inflated, varying greatly in relative height. Sutures distinct, flush with surface. Wall calcareous, smooth, finely perforate. Apertural terminal, on a cylindrical neck, with a lip.

*Distribution at Sites 502A and 503B.* – *Siphonodosaria consobrina* is relatively common with a scattered distribution at Site 502A. It is rare with few occurrences at Site 503B.

## *Siphonodosaria lepidula* (Schwager, 1866)

Fig. 18I

*Synonymy.* – □1866 *Nodosaria lepidula* n.sp. – Schwager, p. 210, Pl. 5:27–28. □1917b *Nodosaria lepidula* Schwager var. *hispidula* n.subsp. – Cushman, p. 654. □1921 *Nodosaria lepidula* Schwager – Cushman, p. 203, Pl. 36:6. □1921 *Nodosaria lepidula* Schwager var. *hispidula* n.subsp. – Cushman, pp. 203–204, Pl. 36:7. □1923 *Nodosaria antillea* n.sp. – Cushman, p. 91, Pl. 14:9. □1934 *Nodogenerina lepidula* (Schwager) – Cushman, p. 122, Pl. 14:15–16. □1939b *Ellipsonodosaria lepidula* (Schwager) – Cushman, p. 150. □1950 *Siphonodosaria lepidula* (Schwager) – Thalmann, p. 12. □1954 *Siphonodosaria lepidula* (Schwager) – Cushman *et al.*, p. 356, Pl. 88:27–28. □1960 *Stilostomella antillea* (Cushman) – Barker, p. 158, Pl. 76:9?, 10. □1971 *Stilostomella antillea* (Cushman) – Schnitker, p. 212, Pl. 5:3a–b. □1980 *Stilostomella lepidula* (Schwager) – Keller, Pl. 1:7. □1984a *Stilostomella lepidula* (Schwager) – Boersma, Pl. 7:9. □1984b *Stilostomella lepidula* (Schwager) – Boersma, Pl. 2:13–14. □1985 *Stilostomella antillea* (Cushman) – Hermelin & Scott, p. 217, Pl. 5:5. □1985 *Stilostomella lepidula* (Schwager) – Kohl, pp. 64–64, Pl. 19:4a–d. □1985 *Stilostomella lepidula* (Schwager) – Thomas, p. 678, Pl. 14:8. □1986 *Stilostomella lepidula* (Schwager) – Boersma, p. 990, Pl. 16:1–2. □1986 *Stilostomella lepidula* (Schwager) trans. *Siphonodosaria insecta* (Schwager) – Boersma, p. 990, Pl. 16:3–4. □1986 *Siphonodosaria insecta* (Schwager) – Boersma, p. 990, Pl.

16:5–6. □1989 *Siphonodosaria lepidula* (Schwager) – Hermelin, pp. 61–62, Pl. 11:8–9.

*Description.* – Test free, uniserial, slender; initial end rounded, sometimes with a short spine. Chambers inflated, subglobular, increasing fairly rapidly in size, six to eight in the adult, strongly embracing in the early portion of test, more remote in the later portion. Sutures distinct, depressed. Wall calcareous, hyaline, finely perforate, ornamented by a series of spines encircling the lower half of each chamber; spines in single or multiple rows; the upper part of the chambers are covered with very fine spines. Aperture terminal, on a distinct neck, which is ornamented with a collar of short spines, a T- shaped tooth projects into the apertural opening from the phialine lip.

*Remarks.* – The intraspecific variation of *S. lepidula* is great with regard to the amount of alignment of the spines and the degree of spinosity. As a result of this *S. lepidula* has been referred to by many different names, specific as well as generic. I agree with Hermelin (1989) that several of the variants of this species should probably be regarded as junior synonyms of *S. lepidula*.

*Ecology.* – In the eastern Pacific this species is found in the lower middle bathyal zone (Ingle, 1980).

*Distribution at Sites 502A and 503B.* – *Siphonodosaria lepidula* occurs in several samples at both Site 502A and Site 503B, but in relatively low abundances.

# Family Buliminidae Jones, 1875

# Subfamily Buliminidae Jones, 1875

# Genus *Bulimina* d'Orbigny, 1826

## *Bulimina alazanensis* Cushman, 1927

Fig. 18J–K

*Synonymy.* – □1927b *Bulimina alazanensis* n.sp. – Cushman, p. 161, Pl. 25:4. □1949 *Bulimina alazanensis* Cushman, 1927b – Bermúdez, p. 180, Pl. 12:1. □1951 *Bulimina alazanensis* Cushman, 1927b – Phleger & Parker p. 16, Pl. 7:24. □1953 *Bulimina alazanensis* Cushman, 1927b – Phleger *et al.*, p. 32, Pl. 6:23. □1965 *Bulimina alazanensis* Cushman, 1927b – Souaya, p. 314, Pl. 2:9. □1966 *Bulimina alazanensis* Cushman, 1927b – Belford, p. 62, Pl. 5:9–11. □1978 *Bulimina alazanensis* Cushman, 1927b – Wright, p. 712, Pl. 3:5–6. □1984a *Bulimina alazanensis* Cushman, 1927b – Boersma, Pl. 3:4. □1984b *Bulimina alazanensis* Cushman, 1927b – Boersma, p. 663. □1985 *Bulimina alazanensis* Cushman, 1927b – Kohl, p. 66, Pl.

20:2. □1985 *Bulimina alazanensis* Cushman, 1927b – Thomas, Pl. 2:6.

*Description.* – Test free, small, triserial, length almost twice the width, widest near apertural end, rounded in cross-section; initial end acute. Chambers slightly inflated, increasing gradually in size. Sutures indistinct. Wall calcareous, hyaline, coarsely perforate, ornamented by 15 or more costae that are running from the proloculus to the base of the last two chambers. Aperture an elongate slit with toothplate, located in a depression of the ultimate chamber.

*Remarks.* – Considerable confusion exists between *B. alazanensis* and *B. rostrata* Brady. *Bulimina alazanensis* has a reticulate pattern of costae, whereas in *B. rostrata* the costae do not touch. *Bulimina semicostata* resembles *B. alazanensis* closely, but the costae do not cover last chamber.

*Ecology.* – *Bulimina alazanensis* is reported from deeper bathyal water in the Gulf of Mexico (Pflum & Frerichs 1976). According to Rathburn & Corliss (1994), the genus *Bulimina* is restricted primarily to sediments with high organic-carbon and low oxygen conditions in the Sulu Sea.

*Distribution at Sites 502A and 503B.* – *Bulimina alazanensis* is the most abundant species of this genus at Site 502A, with relative abundances between 0 and 5%. This species occurs throughout the studied sequence. At Site 503B *B. alazanensis* is a rare species with scattered occurrences above 62.77 m.

## *Bulimina marginata* d'Orbigny, 1826

Fig. 18L

*Synonymy.* – □1826 *Bulimina marginata* n.sp. – d'Orbigny, p. 269. □1949 *Bulimina marginata* d'Orbigny, 1826 – Bermúdez, p. 182, Pl. 12:11. □1951 *Bulimina marginata* d'Orbigny, 1826 – Hofker, p. 154, Text-figs. 95–96. □1951 *Bulimina marginata* d'Orbigny, 1826 – Phleger & Parker, p. 16, Pl. 7:27–28. □1953 *Bulimina marginata* d'Orbigny, 1826 – Phleger *et al.*, p. 33, Pl. 6:25–26. □1957 *Bulimina marginata* d'Orbigny, 1826 – Todd & Brönnimann, p. 32, Pl. 8:4–5. □1960 *Bulimina marginata* d'Orbigny, 1826 – Barker, p. 104, Pl. 51:3–5. □1964 *Bulimina marginata* d'Orbigny, 1826 – Leroy, p. 30, Pl. 11:2. □1971 *Bulimina marginata* d'Orbigny, 1826 – Murray, p. 119, Pl. 49:1–7. □1985 *Bulimina marginata* d'Orbigny, 1826 – Kohl, p. 66, Pl. 20:3. □1986 *Bulimina marginata* d'Orbigny, 1826 – Van Morkhoven *et al.*, pp. 18–21, Pl. 2:1a–b.

*Description.* – Test free, triserial, length 1.5 times the width, widest near middle of test, ovate in cross-section, initial end acute. Chambers inflated, undercut at the margin, rapidly increasing in size as added. Sutures distinct, depressed, generally obscured by ornamentation on early portion, oblique in the later portion. Wall calcareous, hyaline, finely perforate, smooth; margins of chambers ornamented with short downward-projecting spines decreasing in number on later chambers. Aperture a loop-shaped slit with toothplate in the apertural face at the inner margin of the ultimate chamber, surrounded by a raised lip.

*Ecology.* – According to Phleger (1951) and Parker (1954) *Bulimina marginata* range from water depths less than 3 m down to the upper bathyal zone. In the Gulf of Mexico, Plfum & Frerichs (1976) reported this species to be restricted to fine-grained substrate and stenohaline conditions at depths of <30 m to approximately 480 m. According to Bandy *et al.* (1965), *B. marginata* is rarely encountered in deeper bathyal assemblages and is found in lagoonal areas as well as in the lower neritic and upper bathyal zones off California. According to Corliss & Emerson (1990), *Bulimina* is a shallow infaunal genus, commonly found within low-oxygen environments where sediments are relatively high in organic carbon; and *B. marginata* occur in its greatest abundances at the outer shelf and upper slope depths along the continental margin of the northeastern United States (Miller & Lohmann 1982) and in the Gulf of Mexico (Poag 1981). This relationship is not observed in the South China and Sulu Seas, where abundant buliminids do not occur within the oxygen minimum zone (Miao & Thunell 1993). Miao & Thunell (1993) found instead that buliminids are most abundant in sediments where bottom waters are well-oxygenated and organic-carbon contents are high in the South China Sea. This further supports the concept that an infaunal habitat is mainly controlled by organic-carbon content in the sediment and not by oxygen concentration in overlying bottom water as proposed by Corliss (1985) and Corliss & Emerson (1990). Collins (1990) found populations of *B. marginata* predominantly dextrally coiled in waters with warm temperatures, but there was no correlation between cold temperatures and the predominantly sinistral populations in the Gulf of Maine and Gulf of Mexico.

*Distribution at Sites 502A and 503B.* – *Bulimina marginata* is represented by only two specimens at Site 503B (87.44m). At Site 502A it is absent.

## *Bulimina translucens* Parker, 1953

Fig. 18M

*Synonymy.* – □1953 *Bulimina translucens* n.sp. – Parker *in* Phleger *et al.*, p. 33, Pl. 6:30–31. □1978 *Bulimina translucens* Parker, 1953 – Boltovskoy, p. 154, Pl. 2:8.

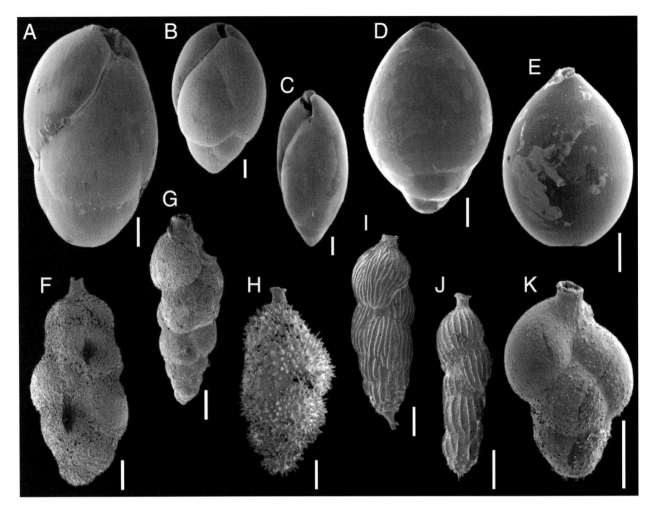

*Fig. 19.* Scale bar 100 μm. □A, B. *Globobulimina affinis* (d'Orbigny), A Side view, Hole 503B, 94.67 m; B Side view, Hole 503B, 109.93 m. □C. *Globob-ulimina auriculata* (Bailey), Side view, Hole 503B, 56.89 m. □D. *Globobulimina pacifica* Cushman, Side view, Hole 502A, 48.80 m. □E. *Globobulimina saubriguensis* (Rahaghi), Side view, Hole 502A, 76.72 m. □F. *Uvigerina auberiana* d'Orbigny, Side view, Hole 503B, 59.00 m. □G. *Uvigerina canariensis* d'Orbigny, Side view, Hole 503B, 77.35 m. □H. *Uvigerina hispida* Schwager, Side view, Hole 503B, 41.08 m. □I, J. *Uvigerina hollicki* Thalmann; I, Side view, Hole 502A, 77.69 m; □J. Side view, Hole 502A, 77.69 m. □K. *Uvigerina mantaensis* Cushman & Edwards, Side view, Hole 502A, 90.94 m.

*Description.* – Test free, small, slender, gradually tapering from the widest point at the midpoint of the last-formed whorl; megalospheric form about twice as long as broad, microspheric form three times as long as wide, initial end rounded, the microspheric form with 6–7 whorls, the megalospheric about 5. Chambers distinct, slightly inflated, increasing gradually in size as added. Sutures distinct, slightly depressed. Wall calcareous, smooth, finely perforate, translucent, the lower part of the test often, not always, decorated by a few low, curving costae which extend as much as halfway up the test. Aperture moderately broad, loopshaped, located above the junction of the second and third chambers, near the apex of the test.

*Remarks.* – *Bulimina translucens* closely resembles *Bulimina thanetensis* (Cushman & Parker) but is much more finely perforate and is usually costate on the lower part of the test.

*Distribution at Sites 502A and 503B.* – *Bulimina translucens* is a very rare species at Site 503B with only a few occurrences above 66.75 m. It is absent at Site 502A.

## Genus *Globobulimina* Cushman, 1927

## *Globobulimina affinis* (d'Orbigny, 1839)

Fig. 19A–B

*Synonymy.* – □  1839a *Bulimina affinis* sp.nov. – d'Orbigny, p. 105, Pl. 2:25–26. □1945 *Bulimina (Desinob-ulimina) illingi* Cushman & Stainforth, p. 41, Pl. 6:7. □ 1951 *Bulimina affinis* d'Orbigny, 1839a – Phleger & Parker, 1951, p. 15, Pl. 7:21–22. □1954 *Globobulimina affinis* (d'Orbigny) – Parker, Pl. 6:25, Pl. 7:1–2. □1958

*Globobulimina affinis* (d'Orbigny) – Parker, p. 262, Pl. 2:24 (not Pl. 2:25). □1960 *Bulimina pupoides* d'Orbigny? – Barker, p. 102, Pl. 50:14. □1964 *Globobulimina affinis* (d'Orbigny) – Akers & Dorman, p. 35, Pl. 7:13. □ 1964 *Bulimina affinis* d'Orbigny, 1839a – Smith, p. B31–B32, Pl. 2:2–3. □1978 *Globobulimina affinis* (d'Orbigny) – Wright, p. 714. □1990 *Globobulimina auriculata* (Bailey) – Thomas *et al.*, Pl. 6:3–4.

*Description.* – Test free, triserial, pyriform in side view, usually broadest at the middle of the test, ovate in cross-section. Chambers distinct, the last three making up about three fourths of the exterior. Sutures distinct, very weakly depressed. Wall calcareous, smooth, finely perforate, no spines or denticles. Aperture open or nearly so, the tip of the tongue fan- shaped, the free border of the aperture with a weakly marked collar.

*Ecology.* – In the Gulf of Mexico, *G. affinis* is found to have a depth range from the lower neritic zone to abyssal water depth (Pflum & Frerichs 1976). According to Rathburn & Corliss (1994) the genus *Globobulimina* is generally adapted to high organic-carbon and low-oxygen conditions of deep infaunal microhabitats. The deep infaunal occurrence of *Globobulimina* species have been reported by a number of studies (for a review, see Sen Gupta & Machain-Castillo 1993).

*Distribution at Sites 502A and 503B.* – *Globobulimina affinis* is represented by a single specimen at Site 502A (72.39 m). It is rare with few occurrences at Site 503B.

## *Globobulimina auriculata* (Bailey, 1851)

Fig. 19C

*Synonymy.* – □1851 *Bulimina auriculata* n.sp. – Bailey, p. 12, Pl. 1:25–27. □1947 *Globobulimina auriculata gullmariensis* Höglund, p. 252 (pp. 237–254), Pl. 20:6; Pl. 21:5; Pl. 22:6; Text-figs. 258–265, 268–269; 271. □1948 *Bulimina auriculata* Bailey, 1851 – Cushman & McCulloch, p. 249, Pl. 31:4. □1953 *Globobulimina affinis* (d'Orbigny) – Phleger *et al.*, p. 34, Pl. 6:32. □1958 *Globobulimina affinis* (d'Orbigny) – Parker, Pl. 2:25. □1964 *Bulimina auriculata* Bailey, 1851 – Smith, B32, Pl. 2:4a–b. □1980 *Globobulimina auriculata* (Bailey) – Keller, Pl. 2:1. p. 150, Pl. 571:4–7. □1988 *Globobulimina auriculata* (Bailey) – Loeblich & Tappan, p. 150, Pl. 571:4–7.

*Description.* – Test free, ovate to fusiform in lateral view, circular in cross-section, usually broadest at or somewhat below the middle of the test. Chambers distinct, those in the last whorl composing the greatest part (about four-fifths or more) of the test. Sutures distinct, in the initial part slightly depressed, later almost flush with surface. Wall calcareous, smooth, finely perforate. Aperture terminal, loop-shaped, the tip of the tongue fan-shaped, the free border of the aperture with a slight collar.

*Remarks.* – *Globobulimina auriculata* differs from *Globobulimina affinis* by its more loop-shaped and terminal aperture, as well as the less dense perforate wall.

*Distribution at Sites 502A and 503B.* – *Globobulimina auriculata* is a rare species with scattered occurrences at both investigated sites.

## *Globobulimina pacifica* Cushman, 1927

Fig. 19D

*Synonymy.* – □1927a *Globobulimina pacifica* n.sp. – Cushman, p. 67, Pl. 14:12a–b. □1944 *Globobulimina pacifica* Cushman, 1927a – Franklin, p. 314, Pl. 46:19. □1944a *Globobulimina pacifica* Cushman, 1927a – Leroy, p. 27, Pl. 5:12. □1947 *Globobulimina pacifica* Cushman, 1927a – Cushman & Parker, p. 134, Pl. 29:37. □1958 *Globobulimina pacifica* Cushman, 1927a – Asano, Pl. 11, Pl. 2:10a–b. □1960 *Globobulimina pacifica* Cushman, 1927a – Barker, p. 102, Pl. 50:7–10. □1964 *Globobulimina pacifica* Cushman, 1927a – Leroy, p. 30, Pl. 14:3. □1985 *Globobulimina pacifica* Cushman, 1927a – Kohl, p. 67, Pl. 21:1.

*Description.* – Test free, triserial, pyriform in side view, broadest near base of the test, decreasing in width toward the apertural end, ovate in cross-section. Chambers inflated, strongly overlapping, last whorl only visible in side view, earlier chambers seen in basal view. Sutures distinct, flush with surface. Wall calcareous, smooth, finely perforate. Aperture loop- shaped, terminal, with a broad projecting toothplate.

*Remarks.* – *Globobulimina pacifica* Cushman var. *scalprata* Cushman & Todd (1945), differs from *G. pacifica* in having many fine, longitudinal striae.

*Distribution at Sites 502A and 503B.* – *Globobulimina pacifica* is a rare species with scattered occurrences at Site 502A. It is absent at Site 503B.

## *Globobulimina saubriguensis* (Rahaghi, 1977)

Fig. 19E

*Synonymy.* – □1977 *Cuvillierella saubriguensis* – Rahaghi, p. 166. □1988 *Globobulimina saubriguensis* (Rahaghi) – Loeblich & Tappan, p. 521, Pl. 571:17–19.

*Description.* – Test free, round to ovate in outline, circular in cross-section, triserial, broadest at the middle of the test. Chambers inflated, strongly overlapping; later chambers may partially or completely overlap the proceeding ones. Sutures oblique, slightly depressed. Wall calcareous, surface smooth, finely perforate. Aperture terminal, loop-shaped.

*Remarks.* – For the genus *Globobulimina* the apertural characters appear to be more important generically than is either the degree of chamber overlap or the occasional closing of the base of the aperture. According to Loeblich & Tappan (1988) *Cuvillierella* should be regarded as a synonym of *Globobulimina*.

*Distribution at Sites 502A and 503B.* – *Globobulimina saubriguensis* is rare with few occurrences at both Site 502A and Site 503B.

# Family Uvigerinidae Haeckel, 1894

# Genus *Uvigerina* d'Orbigny, 1826

Deep-sea uvigerinids show variation in the number and distribution of costae and spines, as well as the shape and size of the test. This variation may be the result of phenotypic variation within one species, or the presence of a number of species.

## *Uvigerina auberiana* d'Orbigny, 1839

Fig. 19F

*Synonymy.* – ☐1839a *Uvigerina auberiana* n.sp. – d'Orbigny, p. 106, Pl. 2:23–24. ☐1866 *Uvigerina proboscidea* n.sp. – Schwager, p. 250, Pl. 7:96. ☐1884 *Uvigerina asperula* Czjzek, var. *ampullacea* Brady, p. 579, Pl. 74:10–11. ☐1913 *Uvigerina proboscidea* Schwager, 1866 – Cushman, p. 94, Pl. 42:2a–b. ☐1913 *Uvigerina ampullacea* Brady – Cushman, p. 102, Pl. 42:3a–b. ☐1921 *Uvigerina ampullacea* Brady – Cushman, pp. 274–275, Pl. 55:7. ☐1923 *Uvigerina auberiana* d'Orbigny, 1839a – Cushman, p. 163, Pl. 42:3–4. ☐1933b *Uvigerina proboscidea* Schwager var. *vadescens* – Cushman, p. 85, Pl. 8:14–15. ☐1942 *Uvigerina proboscidea* Schwager var. *vadescens* – Cushman, pp. 50–51, Pl. 14:5–9. ☐1942 *Uvigerina ampullacea* Brady – Cushman, pp. 46–48, Pl. 13:2–6. ☐1942 *Uvigerina proboscidea* Schwager, 1866 – Cushman, pp. 49–50, Pl. 14:1–4. ☐1945 *Uvigerina proboscidea* Schwager, 1866 – Cushman & Todd, p. 50, Pl. 7:28a–b. ☐1951 *Neouvigerina ampullacea* (Brady) – Hofker, pp. 213–216, Figs. 140–142. ☐1953 *Uvigerina auberiana* d'Orbigny, 1839a– Phleger *et al.*, p. 37, Pl. 7:30–35. ☐1964 *Uvigerina proboscidea* Schwager var. *vadescens* – Leroy, p. 35, Pl. 3:38. ☐1964 *Uvigerina proboscidea* Schwager, 1866 – Leroy, p. 35, Pl. 16:8. ☐1973 *Uvigerina vadescens* (Cushman) – Douglas, Pl. 8:7. ☐1973 *Uvigerina proboscidea* Schwager,

1866 – Douglas, Pl. 8:8. ☐1974 *Uvigerina auberiana* d'Orbigny, 1839a – Leroy & Levinson, p. 9, Pl. 5:12. ☐1978 *Uvigerina proboscidea* Schwager, 1866 – Boltovskoy, Pl. 8:22–23. ☐1978 *Uvigerina auberiana* d'Orbigny, 1839a – Lohmann, p. 26, Pl. 4:16. ☐1980 *Uvigerina auberiana* d'Orbigny, 1839a – Ingle *et al.*, Pl. 6:1. ☐1981 *Siphouvigerina interrupta* (Brady) – Burke, Pl. 1:16 (not *Uvigerina interrupta* Brady, 1879). ☐1981 *Uvigerina auberiana* d'Orbigny, 1839a – Resig, Pl. 2:2. ☐1984a *Uvigerina auberiana* d'Orbigny, 1839a – Boersma, Pl. 7:8. ☐1984b *Uvigerina auberiana* d'Orbigny, 1839a – Boersma, Pl. 3:4. ☐1985 *Uvigerina auberiana* d'Orbigny, 1839a – Hermelin & Scott, p. 218, Pl. 3:8. ☐1986 *Uvigerina auberiana* d'Orbigny, 1839a – Boersma, Pl. 20:1. ☐1984a *Uvigerina proboscidea* Schwager, 1866 – Boersma, Pl. 8:3. ☐1984b *Uvigerina proboscidea* Schwager, 1866 – Boersma, Pl. 3:5, 8. ☐1985 *Siphouvigerina auberiana* (d'Orbigny) – Kohl, pp. 70–71, Pls. 22:7–8; 23:1. ☐1986 *Uvigerina proboscidea* Schwager, 1866 – Boersma, Pl. 20:2. ☐1986 *Uvigerina proboscidea* Schwager, 1866 – Kurihara & Kennett, Pl. 3:6. ☐1989 *Uvigerina proboscidea* Schwager, 1866 – Hermelin, p. 64, Pl. 12:4–5. ☐1990 *Uvigerina proboscidea* Schwager, 1866 – Ujiié, p. 32, Pl. 13:10–11. ☐1990 *Uvigerina proboscidea* Schwager var. *vadescens* – Ujiié, p. 32, Pl. 13:9. ☐1990 *Neouvigerina ampullacea* (Brady) – Ujiié, p. 32, Pl. 12:12.

*Description.* – Test free, initial portion triserial, later twisted biserial, elongate, circular in cross-section; initial end blunt and sometimes with small spine. Chambers inflated, narrowing towards the apertural neck. Wall calcareous, hyaline, finely perforate, ornamented with small, sharp spines, randomly arranged over the test.

*Remarks.* – In the literature, numerous small, finely pitted uvigerinid species have been described, and several of them seem to be junior synonyms of *U. auberiana* (Hermelin 1989). Boltovskoy (1978) regarded many different variants as *U. proboscidea*. In addition to the species mentioned in the synonymy list, Hermelin (1989) regarded some other species as junior synonyms of *U. auberiana*, e.g., *Uvigerina asperula* Czjzek, *Uvigerina interrupta* Brady, and *Uvigerina senticosa* Cushman. The species figured by Ujiié (1990) as *Neouvigerina ampullacea* and by Burke (1981) as *Siphouvigerina interrupta* (Brady) are here regarded as synonymous of *U. auberiana*.

*Ecology.* – *Uvigerina auberiana* has been reported from the upper middle to lower bathyal zones off the west coast of Central America (Smith, 1964). In the Gulf of Mexico, Pflum & Frerichs (1976) found *U. auberiana* in the lower neritic zone. In the South China Sea, *U. auberiana* is found with frequencis of 10% during glacial stage and decreasing to 5% during stage 1 (Miao & Thunell 1996).

*Distribution at Sites 502A and 503B.* – *Uvigerina auberi-ana* is represented by scattered occurrences at Site 502A. It is more abundant at Site 503B.

## *Uvigerina canariensis* d'Orbigny, 1839

Fig. 19G

*Synonymy.* – □1839a *Uvigerina canariensis* n.sp. – d'Orbigny, p. 138, Pl. 1:25– 27. □1913 *Uvigerina canar-iensis* d'Orbigny, 1839a – Cushman, p. 92, Pl. 42:6. □1960 *Uvigerina canariensis* d'Orbigny, 1839a – Barker, p. 154, Pl. 74:1–3. □1981 *Uvigerina canariensis* d'Orbigny, 1839a – Cole, p. 91, Pl. 19:44. □1985 *Uvigerina canariensis* d'Orbigny, 1839a – Hermelin & Scott, p. 218, Pl. 3:9.

*Description.* – Test free, elongate, more than twice as long as wide, often with the widest part below mid-line. Sutures indistinct, depressed. Wall calcareous, perforated, finely pitted. Aperture terminal, on a short neck with a collar.

*Distribution at Sites 502A and 503B.* – *Uvigerina canar-iensis* is absent at Site 502A and is very rare with only a few occurrences at Site 503B.

## *Uvigerina hispida* Schwager, 1866

Fig. 19H

*Synonymy.* – □1866 *Uvigerina hispida* n.sp. – Schwager, p. 249, Pl. 7:95. □1964 *Uvigerina hispida* Schwager, 1866 – Leroy, p. 34, Pl. 4:2–3. □1966 *Euuvigerina hispida* (Schwager) – Belford, p. 78, Pl. 7:14–16. □1974 *Uvigerina hispida* Schwager, 1866 – Leroy & Levinson, p. 10, Pl. 5:16–17. □1976 *Uvigerina hispida* Schwager, 1866 – Pflum & Frerichs, Pl. 8:8–10. □1978 *Uvigerina hispida* Schwager, 1866 – Boltovskoy, Pl. 8:12–16. □1980 *Uvige-rina hispida* Schwager, 1866 – Ingle *et al.*, Pl. 8:8. □1984a *Uvigerina hispida* Schwager, 1866 – Boersma, Pl. 5:3. □1984b *Uvigerina hispida* Schwager, 1866 – Boersma, Pl. 3:6–7. □1986 *Uvigerina hispida* Schwager, 1866 – Boersma, Pl. 20:5–6. □1986 *Uvigerina hispida* Schwager, 1866 – Kurihara & Kennett, Pl. 3:7–8. □1986 *Uvigerina hispida* Schwager, 1866 – Van Morkhoven *et al.*, pp. 62–64, Pl. 20:1–4. □1989 *Uvigerina hispida* Schwager, 1866 – Hermelin, p. 65.

*Description.* – Test free, triserial, elongate, length 1.5–2 times the width, circular in cross-section. Chambers inflated, increasing rapidly in size. Sutures indistinct, depressed. Wall calcareous, perforate, ornamented with closely arranged spines. Aperture terminal, at the end of a short neck, with a phialine lip.

*Remarks.* – *Uvigerina hispida* is recognised on the basis of its spines or nodes, and the lack of longitudinal costae.

*Ecology.* – Bandy (1961) reported that *U. hispida* occurs in the upper part of the upper middle bathyal zone in the Gulf of California. Pflum & Frerichs (1976) showed that this species is representative from the abyssal zone to the top of the lower middle bathyal zone in the Gulf of Mex-ico,

*Distribution at Sites 502A and 503B.* – *Uvigerina hispida* is more abundant at Site 503B than at Site 502A, where it is rare with scattered occurrences.

## *Uvigerina hollicki* Thalmann, 1950

Fig. 19I–J

*Synonymy.* – □1923 *Uvigerina peregrina* Cushman var. *bradyana* – Cushman, p. 168, Pl. 42:12 (not *U. bradyana* Fornasini, 1900). □ 1950 *Uvigerina hollicki* nom.nov. – Thalmann, p. 45. □ 1953 *Uvigerina hollicki* Thalmann, 1950 – Phleger *et al.*, Pl. 8:1.

*Description.* – Test free, triserial, 1.5–2 times as long as broad, widest near middle of test. Chambers inflated, embracing. Sutures depressed, indistinct. Wall calcare-ous, perforate, ornamented by numerous bladelike costae on each chamber. Aperture terminal, on a short neck, often with a phialine lip.

*Remarks.* – Like most species of *Uvigerina*, *Uvigerina hol-licki* exhibits a very variable morphology. The morphol-ogy is very similar to that of the *Uvigerina peregrina* group, but the costae are lower and the test often larger in size. Today, this species is regarded by many workers as an ecophenotypic variant of *U. peregrina*, but it is here recog-nised as a separate species.

*Ecology.* – Studies by Phleger *et al.* (1953) in the North Atlantic indicate that this species replaces *U. peregrina* in deeper waters.

*Distribution at Sites 502A and 503B.* – *Uvigerina hollicki* is a common species at Site 502A, where it occurs mainly between 71.42–86.98 m; above and below this interval it has more scattered occurrences. It is very rare at Site 503B.

## *Uvigerina mantaensis* Cushman & Edwards, 1938

Fig. 19K

*Synonymy.* – □1929 *Uvigerina proboscidea* Schwager – Galloway & Morray, p. 39, Pl. 6:4 (not *U. proboscidea* Schwager, 1866). □1938 *Uvigerina mantaensis* sp. nov. – Cushman & Edwards, p. 84, Pl. 14:8. □1941 *Uvigerina mantaensis* Cushman & Edwards, 1938 – Galloway & Heminway, p. 430, Pl. 33:7. □1945 *Uvigerina mantaensis* Cushman & Edwards, 1938 – Cushman & Stainforth, p.

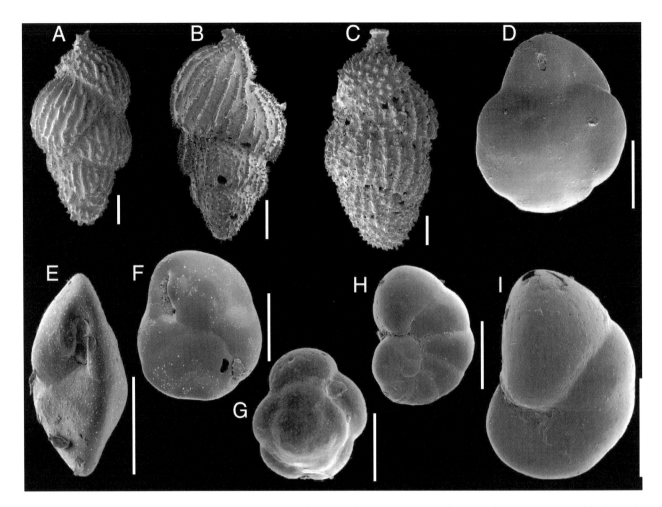

*Fig. 20.* Scale bar 100 μm. □A–C. *Uvigerina peregrina* Cushman; A, Side view, Hole 502A, 89.58 m; B, Side view, Hole 502A, 89.58 m; C, Side view, Hole 502A, 89.58 m. □D–F, *Epistominella exigua* (Brady); D, Spiral view, Hole 502A, 41.62 m; E, Edge view, Hole 502A, 41.62 m; F, Umbilical view, Hole 502A, 41.62 m. □G. *Eponides bradyi* Earland, Spiral view, Hole 502A, 83.99 m. □H, I. *Valvulineria humilis* (d'Orbigny); H, Spiral view, Hole 502A, 97.80 m; I, Umbilical view, Hole 502A, 97.80 m.

47, Pl. 7:17. □1947 *Uvigerina mantaensis* Cushman & Edwards, 1938 – Cushman & Stone, p. 17, Pl. 2:26. □1949 *Uvigerina mantaensis* Cushman & Edwards, 1938 – Bermúdez, p. 207, Pl. 13:48. □1958 *Uvigerina mantaensis* Cushman & Edwards, 1938 – Becker & Dusenbury, p. 33, Pl. 4:22. □1976 *Uvigerina mantaensis* Cushman & Edwards, 1938 – Todd & Low, p. 25, Pl. 4:5. □1985 *Uvigerina mantaensis* Cushman & Edwards, 1938 – Kohl, p. 73, Pl. 24:4.

*Description.* – Test free, triserial, 1.5 times as long as wide, greatest width near the middle of the test, ovate in cross-section, initial end blunt. Chambers few, inflated. Sutures indistinct in early stage, fairly distinct in later stages, slightly depressed. Wall calcareous, perforate, ornamented with fine spines on some intervals. Aperture terminal, on a short hispid neck, with a phialine lip and toothplate.

*Distribution at Sites 502A and 503B.* – *Uvigerina mantensis* is a rare species with few occurrences at both Site 502A and Site 503B.

## *Uvigerina peregrina* Cushman, 1923

Fig. 20A–C

*Synonymy.* – □1923 *Uvigerina peregrina* sp.nov. – Cushman, p. 166, Pl. 42:7–10. □1927 *Uvigerina peregrina* Cushman, 1923 – Galloway & Wissler, p. 76, Pl. 12:1, 2a–b. □1951 *Uvigerina peregrina* Cushman, 1923 – Phleger & Parker (part), p. 18, Pl. 8:22, 24, 25 (not Pl. 8:26). □1958 *Uvigerina peregrina* Cushman, 1923 – Parker, p. 263, Pl. 2:37–38. □1964 *Uvigerina peregrina* Cushman, 1923 – Akers & Dorman, p. 58, Pl. 9:6. □1964 *Uvigerina peregrina* Cushman, 1923 – Smith, p. 84, Pl. 2:15–16. □1971 *Uvigerina peregrina* Cushman, 1923 – Murray, p. 121, Pl.

50:1–7. □1973 *Uvigerina peregrina* Cushman, 1923– Douglas, Pl. 8:4–6, 9. □1974 *Uvigerina peregrina* Cushman, 1923 – Leroy & Levinson, p. 10, Pl. 5:18. □1978 *Uvigerina peregrina* Cushman, 1923 – Boltovskoy, Pl. 8:4–5. □1978 *Uvigerina peregrina* Cushman, 1923 – Lohmann, p. 26, Pl. 4:14–15. □1981 *Uvigerina peregrina* Cushman, 1923 – Cole, p. 92, Pl. 10:9. □1984a *Uvigerina peregrina* Cushman, 1923 – Boersma, Pl. 7:6. □1984b *Uvigerina peregrina* Cushman, 1923 – Boersma, Pl. 3:3. □1985 *Uvigerina peregrina* Cushman, 1923 – Hermelin & Scott, p. 218, Pl. 3:10. □1985 *Uvigerina peregrina* Cushman, 1923 – Kohl, pp. 73–74, Pl. 24:7. □1985 *Uvigerina peregrina* Cushman, 1923 – Mead, p. 229, Pl. 1:7–10. □1986 *Uvigerina peregrina* Cushman, 1923 – Kurihara & Kennett, Pl. 3:1–3. □1989 *Uvigerina peregrina* Cushman, 1923– Hermelin, pp. 66–67, Pl. 12:6–8. □1990 *Uvigerina peregrina* Cushman, 1923 – Thomas *et al.*, Pl. 6:5–6.

*Description.* – Test free, triserial, elongate, about 2.5 times as long as broad, widest near middle of test. Chambers inflated, embracing. Sutures indistinct, depressed. Wall calcareous, perforate, ornamented by numerous high bladelike costae on each chamber; costae break up into spinose or irregular short segments toward the apertural end. Aperture circular at end of short neck, with a phialine lip.

*Remarks.* – *Uvigerina peregrina* is similar in some aspects to *Uvigerina mediterranea* but it has more chambers, which is less inflated and also ornamented by a larger number of smaller, discontinuous costae.

*Ecology.* – In many regions, *U. peregrina* is presently the dominant taxon at intermediate water depths (Pflum & Frerichs 1976; Streeter & Shackleton 1979; Schnitker 1979, 1980; Qvale & van Weering 1985; Gupta & Srinivasan 1990), although this taxon can also be found at abyssal depths (Corliss 1979). Earlier studies often associated *U. peregrina* with low-oxygen bottom waters (Pflum & Frerichs 1976; Lohmann 1978; Schnitker 1979; Streeter & Shackleton 1979). However, more recent studies indicate that *Uvigerina* in general may vary independently of the dissolved-oxygen content of the bottom water (Miller & Lohmann 1982; Ross & Kennett 1983; Woodruff 1985; Corliss *et al.* 1986; Mead & Kennett 1987; Gupta & Srinivasan 1992; Herguera 1992). Studies of living benthic foraminifera (Corliss & Emerson 1990; Rathbun & Corliss 1994) showed that some recent *Uvigerina* spp. are shallow-infaunal taxa and are more influenced by sediment properties than bottom-water content. High *Uvigerina* abundance are generally associated with high sedimentary organic-carbon content (Miller & Lohmann 1982; Woodruff 1985; Corliss *et al.* 1986; Boyle 1990;

Gupta & Srinivasan 1992; Burke *et al.* 1993), and this has been observed in the modern sediments from the South China Sea and Sulu Sea (Tappa 1992; Miao & Thunell, 1993; Rathbun & Corliss 1994). According to Miao & Thunell (1993), *Uvigerina* is the most important benthic foraminifer species in surface sediments from water depths above 1,500 m in the South China Sea, where pore-water oxygen penetration depth is shallow and organic-carbon contents are high. However, Joyce & Williams (1986) found no obvious correlation between variations in the distribution of *U. peregrina* with changes in the content of sedimentary organic carbon in the Gulf of Mexico.

Studies by Schnitker (1974), Streeter & Shackleton (1979), Corliss (1982), and Herguera (1992) suggest that *Uvigerina* species, in general, are more abundant during glacial periods throughout the global ocean. In the North Atlantic, *Uvigerina* is found abundant throughout glacial times at water depths from 2,500 to 4,000 m, and this faunal increase is attributed to a reduction of NADW formation during glacial periods (Streeter & Lavery 1982; Streeter & Shackleton 1979). Berger & Killingley (1982), Corliss *et al.* (1986), and Pedersen *et al.* (1988) found that *Uvigerina* was more abundant in the Pacific and Indian Oceans during glacial times, which they attributed to higher organic-carbon fluxes and increased food availability to the sea-floor at these periods. *Uvigerina* was abundant down to 3,000 m during the last glacial in the South China Sea, which is most likely a response to the higher organic-carbon content of these sediments (Thunell *et al.* 1992).

*Uvigerina peregrina* is found in the lower bathyal zone of the South Atlantic (Mead 1985). Matoba & Yamaguchi (1982) reported this species from the bathyal zone in the Gulf of California. In contrast, off the west coast of Central America it occurs in the middle bathyal zone (Smith 1964). According to Corliss (1991), *U. peregrina* is characterized as a shallow-infaunal species and found inhabiting surficial sediments down to 2–2.5 cm on the Nova Scotian margin, while most *Uvigerina* specimens from areas outside of the North Atlantic were obtained from the the upper 1.5–2 cm (Mackensen & Douglas 1989; McCorkle *et al.* 1990). According to Zahn *et al.* (1986), the $^{13}C$ composition of the infaunal species *U. peregrina* is correlated with the organic-carbon flux and, therefore, has a lighter carbon isotopic composition than epifaunal species.

*Distribution at Sites 502A and 503B.* – *Uvigerina peregrina* is a common species at Site 502A with absolute abundances ranging between 0 and 114. It is rare with scattered occurrences at Site 503B.

## Superfamily Discorbacea Ehrenberg, 1838

## Family Discorbidae Ehrenberg, 1838

## Subfamily Discorbinae Ehrenberg, 1838

## Genus *Epistominella* Husezima & Maruhasi, 1944

### *Epistominella exigua* (Brady, 1884)

Fig. 20D–F

*Synonymy.* – □1884 *Pulvinulina exigua* n.sp. – Brady, p. 696, Pl. 103:13–14. □1921 *Pulvinulina exigua* Brady, 1884 – Cushman, p. 340, Pl. 68:3a–c. □1931 *Pseudoparrella exigua* (Brady) – Cushman & Parker, p. 21. □1931 *Eponides exigua* (Brady) – Cushman, pp. 44–45, Pl. 10:1–2. □1931 *Pulvinulinella exigua* (Brady) – Wiesner, p. 121. □1937 *Eponides exiguus* (Brady) – Chapman & Parr, p. 107. □1950 *Pulvinulinella exigua* (Brady) – Parr, p. 361. □1951 *Pseudoparrella exigua* (Brady) – Phleger & Parker, p. 28, Pl. 15:6a–b, 7a–b. □1953 *Epistominella exigua* (Brady) – Phleger *et al.*, p. 43, Pl. 9:35–36. □1954 *Epistominella exigua* (Brady) – Parker, p. 33, Pl. 10:22–25. □1960 *Epistominella exigua* (Brady) – Barker, p. 212, Pl. 103:13–14. □1964 *Epistominella exigua* (Brady) – Smith, p. 43, Pl. 4:6a–b. □1965 *Epistominella exigua* (Brady) – Todd, pp. 30–31, Pl. 10:1. □1976 *Epistominella exigua* (Brady) – Resig, Pl. 3:1. □1978 *Epistominella exigua* (Brady) – Boltovskoy, Pl. 3:37–38. □1979 *Epistominella exigua* (Brady) – Corliss, p. 7, Pl. 2:7–9. □1981 *Epistominella exigua* (Brady) – Burke, Pl. 2:1–2. □1981 *Epistominella exigua* (Brady) – Cole, 1981, p. 95, Pl. 11:2. □1981 *Epistominella exigua* (Brady) – Resig, Pl. 6:6–7. □1984 *Epistominella exigua* (Brady) – Murray, Pl. 2:1–2. □1985 *Epistominella exigua* (Brady) – Hermelin & Scott, p. 208, Pl. 4:1. □1985 *Epistominella exigua* (Brady) – Mead, p. 230, Pl. 2:1–4. □1985 *Epistominella exigua* (Brady) – Thomas, Pl. 13:3–4. □1986 *Epistominella exigua* (Brady) – Kurihara & Kennett, Pl. 3:10–12. □1989 *Epistominella exigua* (Brady) – Hermelin, p. 67. □1990 *Epistominella exigua* (Brady) – Mackensen *et al.*, p. 252, Pl. 7:1–2. □1990 *Epistominella exigua* (Brady) – Thomas *et al.*, p. 227, Pl. 6:8–11. □1990 *Epistominella exigua* (Brady) – Ujiié, p. 32, Pl. 14:1a–c.

*Description.* – Test free, trochospiral; all chambers are visible on the spiral side, except those of the last whorl on umbilical side. Sutures flush with surface, oblique on spiral side, nearly radial on umbilical side. Wall calcareous, hyaline, smooth, finely perforate. Aperture an elongate vertical slit, parallel to the periphery.

*Remarks.* – The size and shape of the *E. exigua* test show little variation. *Epistominella exigua* characterized by small test size, smooth, transparent test walls and epifaunal (spiral) test morphologies and it is mostly underrepresented or missing in sample fractions >125 μm or larger (see Hermelin 1986). This because *E. exigua* is a small species also in its adult stage.

*Ecology.* – *Epistominella exigua* has been found showing periodic peaks in abundance in the fossil record at many open ocean sites. Previously, these peaks were commonly interpreted as an indicator of changes in physicochemical properties of bottom water mass and thus deep-sea circulation, such as 'young' well-oxygenated Northeast Atlantic Deep Water (NEADW) (Streeter 1973; Schnitker 1974, 1979, 1980; Weston & Murray 1984; Gaydyukov & Lukashina 1988; among others). Murray (1988) interpreted the dramatic increase in abundance and distribution of *E. exigua* at the end of the Pliocene in terms of an increased production of NEADW resulting from circulation changes coincident with the onset of Northern Hemisphere glaciation. Increases in the abundance of *E. exigua* (>63 μm and >150 μm) in the early Miocene to early Pliocene of the Pacific Ocean have been linked to the development of bottom-water masses, particularly Pacific Bottom Water (Kurihara & Kennett 1985; Woodruff 1985). *Epistominella exigua* was also considered as an indicator species of Antarctic Bottom Water (AABW) (Douglas & Woodruff 1981), and lower North Atlantic Deep Water (NADW) (Uchio 1960; Burke 1981). In the Gulf of California, this species is reported from the upper middle bathyal zone, where it showed abundances up to 22% of the total fauna (Bandy 1961). It was found as a marker of the IBW (Indian Bottom Water) by Corliss (1979). Mead (1985) reported it from the bathyal zone of the South Atlantic, whereas Smith (1964) found it in the neritic zone off the west coast of Central America. Boltovskoy *et al.* (1980) and Boltovskoy & Totah (1985) showed that *E. exigua* is abundant on the Argentine shelf. Gooday *et al.* (1992) showed that this species is more dependent on the supply of phytodetritus to the abyssal area than on any typical bottom-water masses. Similarly, Thomas (1992) suggested that the increase in relative abundance of *E. exigua* following the last deglaciation in the northeast Atlantic was linked to a highly increased input of phytodetritus. Also Mackensen (1992) related an uppermost Miocene and Pliocene assemblage (>125m) characterized by *E. exigua* from the southern Indian Ocean to high biosiliceous sedimentation. Smart *et al.* (1994) proposed that *E. exigua* may represent a proxy for seasonal pulses of phytodetritus originating from surface primary productivity in open ocean eutrophic areas, and thereby as an indicator of relative changes in productivity.

*Distribution at Sites 502A and 503B.* – *Epistominella exigua* is a common species that occurs in most samples at both sites, with absolute abundances of 0–115 individuals per sample at Site 502A and 0–39 individuals per sample at Site 503B.

# Genus *Eponides* Montfort, 1808

## *Eponides bradyi* Earland, 1934

Fig. 20G

*Synonymy.* – □1934 *Eponides bradyi* sp.nov. – Earland, p. 187, Pl. 8:36–38. □1960 *Eponides bradyi* Earland, 1934 – Barker, p. 196, Pl. 95:9–10. □1971 *Eponides bradyi* Earland, 1934 – Murray, p. 173, Pl. 72:1–4. □1981 *Eponides bradyi* Earland, 1934 – Cole, p. 102, Pl. 11:5. □1985 *Eponides bradyi* Earland, 1934 – Hermelin & Scott, p. 208, Pl. 4:5–6.

*Description.* – Test free, biconvex, trochospiral, rounded periphery, about three whorls visible on the spiral side. Chambers broad, limbate, curving into the periphery, 6–8 chambers in the last whorl. Sutures depressed and slightly curved back, umbilical side slightly depressed. Wall calcareous, surface smooth, radial, perforated. Aperture an interiomarginal slit.

*Distribution at Sites 502A and 503B.* – *Eponides bradyi* is a common species at certain depths at both investigated sites; at other depths it is rare with scattered occurrences.

## *Eponides polius* (Phleger & Parker, 1951)

*Synonymy.* – □1951 *Eponides polius* n.sp. – Phleger & Parker, p. 21, Pl. 11:1a–b, 2a–b. □1953 *Eponides polius* Phleger & Parker, 1951 – Phleger *et al.*, p. 41, Pl. 9:3–4. □1978 *Eponides polius* Phleger & Parker, 1951 – Boltovskoy, p. 158, Pl. 4:4–5. □1978 *Eponides polius* Phleger & Parker, 1951 – Wright, p. 714, Pl. 4:13–14. □1990 *Eponides polius* Phleger & Parker, 1951 – Ujiié, p. 34, Pl. 16:1a, c, 2a–c.

*Description.* – Test free, small, biconvex, consisting of three whorls, periphery subacute, slightly lobulate. Chambers distinct, 9–10 in the last-formed whorl, increasing slightly in size as added, slightly inflated on ventral side, last-formed chambers project slightly over umbilicus. Sutures distinct, on spiral side flush with surface, slightly curved, rather sharply angled, on ventral side slightly depressed, almost straight. Wall calcareous, surface smooth, radial, perforated. Aperture interiomarginal narrow slit.

*Remarks.* – *Eponides polius* resembles *Oridosalis umbonatus*, but its size is smaller, the chambers are more numerous, and the sutures are more angled on the spiral side.

*Ecology.* – *Eponides polius* ranges in depth from the lower bathyal zone to about 600 m in the Gulf of Mexico (Pflum & Frerichs 1976).

*Distribution at Sites 502A and 503B.* – *Eponides polius* is a rare species with scattered occurrences at Site 502A. It is a relatively common species at Site 503B with absolute abundances of 0–12 individuals per sample.

# Genus *Valvulineria* Cushman, 1926

## *Valvulineria humilus* (d'Orbigny)

Fig. 20H–I

*Synonymy.* – □1839b *Rosalina araucana* n.sp. – d'Orbigny, p. 44, Pl. 6:16–18. □1951 *Valvulineria* cf. *araucana* (d'Orbigny) – Phleger & Parker, p. 25, Pl. 13:7–8. □1953 *Valvulineria araucana* (d'Orbigny) – Phleger *et al.*, p. 40, Pl. 8:29–30. □1964 *Valvulineria glabra* Cushman – Smith, p. 44, Pl. 5:3. □1984b *Valvulineria humilus* (d'Orbigny) – Boersma, Pl. 4:6–7.

*Description.* – Test free, biconvex, periphery broadly rounded, longer than broad in side view, slightly lobate. Chambers distinct, 7–8 in the last whorl, increasing rapidly in size as added, last few chambers inflated. Wall calcareous, smooth, hyaline, coarsely perforate except for the face of the ultimate chamber and the umbilical flap, which are imperforate. Aperture interiomarginal, umbilical–extraumbilical, partially covered by a large projecting umbilical lip.

*Remarks.* – *Valvulineria humilus* is often common in the finer fraction, particularly in the early Pliocene (Boersma 1984b).

*Ecology.* – *Valvulineria* is in general associated with low-oxygen/high organic-carbon environments (van der Zwaan & Jorissen 1991; Sen Gupta & Machain-Castillo 1993).

*Distribution at Sites 502A and 503B.* – *Valvulineria humilus* is a relatively rare species with scattered occurrences at both Site 502A and 503B.

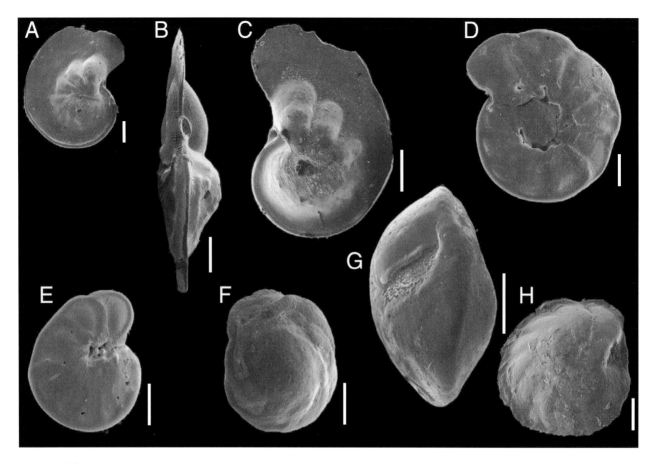

*Fig. 21.* Scale bar 100 μm. □A–C. *Laticarnina pauperata* (Parker & Jones); A, Spiral view, Hole 502A, 66.51 m; B, Edge view, Hole 502A, 66.51 m; C, Umbilical view, Hole 502A, 66.51 m. □D, E. *Hyalinea balthica* (Schroeder); D, Side view, Hole 502A, 48.80 m; E, Opposite side view, Hole 502A, 48.80 m. □F–H *Nuttallides umbonifera* (Cushman); F, Spiral view, Hole 502A, 63.55 m; G, Edge view, Hole 502A, 46.87 m; H, Umbilical view, Hole 502A, 63.55 m.

# Family Laticarininidae Hofker, 1951

# Genus *Laticarinina* Galloway & Wissler, 1927

## *Laticarinina pauperata* (Parker & Jones, 1865)

Fig. 21A–C

*Synonymy.* – □1865 *Pulvinellina repanda* var. *menardii* subvar. *pauperata* – Parker & Jones, p. 395, Pl. 16:50–51. □1931 *Laticarinina pauperata* (Parker & Jones) – Cushman, p. 114, Pls. 20:4a–c; 21:1a–c. □1940 *Laticarinina halophora* (Stache) – Finelay, pp. 467–468. □1941a *Laticarinina pauperata* (Parker & Jones) – Leroy, p. 46, Pl. 2:18–19. □1948 *Laticarinina pauperata* (Parker & Jones) – Renz, p. 143, Pl. 10:4. □1949 *Laticarinina pauperata* (Parker & Jones) –Bermúdez, p. 309, Pl. 23:43–45. □1957 *Laticarinina pauperata* (Parker & Jones) – Agip Miner-

aria, Pl. 50:8a–c. □1960 *Laticarinina halophora* (Stache) – Barker, p. 214, Pl. 104:3–11. □1964 *Laticarinina pauperata* (Parker & Jones) – Leroy, p. 44, Pl. 9:25. □1966 *Laticarinina pauperata* (Parker & Jones) – Belford, p. 92, Pl. 14:9–13. □1981 *Laticarinina halophora* (Stache) – Cole, p. 115, Pl. 14:9. □1985 *Laticarinina halophora* (Stache) – Hermelin & Scott, p. 212, Pl. 4:2–3. □1985 *Laticarinina pauperata* (Parker & Jones) – Kohl, p. 77, Pl. 26:1. □1985 *Laticarinina pauperata* (Parker & Jones) – Thomas, Pl. 11:10. □1985 *Laticarinina pauperata* (Parker & Jones) □1986 *Laticarinina pauperata* (Parker & Jones) – Belanger & Berggren, p. 338, Pl. 3:7a–b. □1986 *Laticarinina pauperata* (Parker & Jones) – Morkhoven *et al.*, pp. 89–90, Pl. 26:1a–c. □1989 *Laticarinina halophora* (Stache) – Hermelin, pp. 68–69. □1990 *Laticarinina pauperata* (Parker & Jones) – Mackensen *et al.*, p. 252, Pl. 7:3. □1990 *Laticarinina pauperata* (Parker & Jones) – Ujiié, p. 33, Pl. 14:3a–c, 4a–c.

*Description.* – Test free, slightly trochoid, compressed, biconvex to planoconvex in edge view, spiral side flat-

*Neogene benthic foraminifera* 59

tened and evolute, ventral side convex and involute. Surrounded with a broad, thin, transparent keel. Sutures radial, depressed on ventral side, flush on spiral side. Wall calcareous, hyaline, smooth, finely perforate. Aperture on the dorsal side at the inner margin of the last-formed chamber.

*Remarks.* – The most distinctive characteristics of *Laticarinina pauperata* are the strongly compressed test and broad, transparent keel.

*Ecology.* – According to Bandy & Rodolfo (1964) *Laticarinina pauperata* is a bathyal–abyssal taxon, which occurs at different depths and in waters of markedly dissimilar character in modern oceans. Its upper depth limit is 1,600 m off southern California (Bandy 1963) and 300 m in the Gulf of Mexico (Parker 1954), where it is approximately twice as abundant in the lower bathyal zone than at middle and upper bathyal depths (Parker 1954; Pflum & Frerichs 1976). This species is considered to be more abundant in regions with high productivity (Woodruff 1985; Woodruff & Savin 1989). Burke *et al.* (1993) reported that *L. pauperata* is more abundant during late transitional time on the Ontong Java Plateau.

From laboratory studies, Weinberg (1991) reported that *L. pauperata* moved in and out of the sediments, suggesting that in nature this species may occupy a range of depths within the sediment.

*Distribution at Sites 502A and 503B.* – *Laticarinina pauperata* is a common species, which is represented in most samples at both sites. Absolute abundances 0–29 individuals per sample at Site 502A and 0–8 individuals per sample at Site 503B.

# Family Planulinidae Bermúdez, 1952

# Genus *Hyalinea* Hofker, 1951

## *Hyalinea balthica* (Schroeder, 1783)

Fig. 21D–E

*Synonymy.* – ☐1783 *Nautilus balthicus* – Schroeder, p. 20, Pl. 1:2. ☐1884 *Operculina ammonoides* Parker & Jones (not Gronovius) – Brady, p. 745, Pl. 92:1–2. ☐1931 *Anomalina balthica* (Schroeder) – Cushman, pp. 108–109, Pl. 19:3a–c. ☐1951 *Hyalinea balthica* (Schroeder) – Hofker, pp. 508–513:345–348. ☐1952 *Hofkerinella balthica* (Schroeder) – Bermúdez, pp. 74–75. ☐1953 *Anomalina balthica* (Schroeder) – Phleger *et al.*, p. 48, Pl. 10:24–25. ☐1960 *Hyalinea balthica* (Schroeder)– Barker, p. 230, Pl. 112:1–2. ☐1964 *Hyalinea balthica* (Schroeder) – Akers & Dorman, pp. 38–39, Pl. 10:18–19. ☐1964 *Hyalinea balthica* (Schroeder) – Feyling–Hanssen, pp. 351–552, Pl.

21:14–16. ☐1964 *Hyalinea balthica* (Schroeder) – Loeblich & Tappan, 1964, pp. C686–C687. ☐1978 *Hyalinea balthica* (Schroeder) – Wright, p. 715, Pl. 5:17. ☐1986 *Hyalinea balthica* (Schroeder) – Van Morkhoven *et al.*, pp. 21–22, Pl. 3:1–3. ☐1988 *Hyalinea balthica* (Schroeder) – Loeblich & Tappan, p. 580, Pl. 632:5–8. ☐1991 *Hyalinea balthica* (Schroeder) – Hermelin, pp. 244–251, Pl. 1:1–16.

*Description.* – Test free, discoidal, slightly trochospiral to nearly planispiral, semievolute on both sides; periphery angled, with broad imperforate keel. Chambers about 8–12 in the last whorl, umbilical margins with a small umbilical flap or folium. Sutures radial, slightly curved, thickened and elevated, nonperforate. Wall calcareous, finely perforate, radial in structure, with septa and marginal keel nonperforate. Aperture a low equatorial and interiomarginal arch with narrow bordering lip, a low slit continuing laterally beneath the folium, apertures remaining open for a few chambers before being closed by lamellar thickening.

*Remarks.* – In Italy, *H. balthica* appeared at the base of the Pleistocene and has been used there to distinguish marine Pleistocene strata from older strata (Alliata 1946, 1947; Coggi & Alliata 1950; Ilacqua 1956). The first recorded occurrence of *H. balthica* is in the Caribbean (Bock 1970). According to Bolli *et al.* (1968), the first appearance of this species was observed to coincide with the extinctions of the calcareous nannofossils *Discoaster brouweri* and *Coccolithus pelagicus*, near the Pliocene–Pleistocene boundary. A study by Hermelin (1991) in the northwest Arabian Sea reveals that the first appearance of *H. balthica* was in the early Pliocene, at about 5.0 Ma.

*Ecology.* – According to Ross (1984), this species is most abundant in deep, cool water masses of the North Atlantic, especially on the eastern side. It has not been reported from Pleistocene sediments of the western Atlantic (Van Morkhoven *et al.* 1986), nor from interglacial sediments of the Gulf of Mexico and the Caribbean (Bock 1970). Colom (1950) recorded it as abundant in the deepest facies he examined, at 300–878 m depth off the west coast of Africa. *Hyalinea balthica* has been reported from deep, cool waters of the Mediterranean. In the Norwegian Channel (Northern North Sea), *H. balthica* occurs most abundantly in sediments with a high content of organic carbon (Qvale & van Weering 1985).

*Distribution at Sites 502A and 503B.* – *Hyalinea balthica* is only found above 107.66 m at Site 502A and 88.41 m at Site 503B. The species is very rare and exhibits scattered occurrences below these levels, and absolute abundances are less than 3 individuals per sample ,except for the sample from 48.80 m at Site 502A, where the absolute abundance is 20.

# Family Epistomariidae Hofker, 1954

## Genus *Nuttallides* Finlay, 1939

### *Nuttallides umbonifera* (Cushman, 1933)

Fig. 21F–H

*Synonymy.* – □1933b *Pulvinulinella umbonifera* n.sp. – Cushman, p. 90, Pl. 9:9a–c. □1934 *Eponides bradyi* sp.nov. – Earland, p. 187, Pl. 8:36–38. □1953 *Epistominella* (?) *umbonifera* (Cushman) – Phleger *et al.*, p. 43, Pl. 9:33–34. □1960 *Eponides bradyi* Earland, 1934 – Barker, p. 196, Pl. 95:9–10. □1965 *Nuttallides umboniferus* (Cushman) – Todd, pp. 29–30, Pl. 11:1. □1971 *Nuttallides umbonifera* (Cushman) – Echols, p. 166, Pl. 14:5a–c. □1973 ?*Epistominella umbonifera* (Cushman) – Streeter, p. 133. □1974 *Osangularia umbonifera* (Cushman) – Schnitker, p. 385. □1975 *Nuttallides umbonifer* (Cushman) – Anderson, p. 88, Pl. □1976 *Osangularia rugosa* (Phleger & Parker) – Pflum & Frerichs, Pl. 7:2–4. 8:14a–c. □1976 *Nuttallides umbonifera* (Cushman) – Resig, Pl. 3:6–7. □1978 "*Epistominella*" *umbonifera* (Cushman) – Lohmann, p. 26, Pl. 3:1–6. □1979 *Epistominella umbonifera* (Cushman) – Corliss, p. 7, Pl. 2:10–12. □1981 *Nuttallides umbonifera* (Cushman) – Burke, Pl. 2:5–6. □1981 *Osangularia rugosa* (Phleger & Parker) – Cole, p. 114, Pl. 20:12–13. □1981 *Nuttallides umbonifera* (Cushman) – Resig, Pl. 6:13–15. □1984a *Nuttallides umbonifera* (Cushman) – Boersma, Pl. 6:6, Pl. 7:12. □1985 *Nuttallides umbonifera* (Cushman) – Hermelin & Scott, p. 214, Pl. 5:11–13. □1985 *Nuttallides umbonifera* (Cushman) – Mead, pp. 230–232, Pl. 2:6–7. □1985 *Nuttallides umbonifera* (Cushman) – Thomas, Pl. 13:1–2. □1989 *Nuttallides umbonifera* (Cushman) – Hermelin, p. 69, Pl. 12:15–17. □1990 *Nuttallides umbonifer* (Cushman) – Mackensen *et al.*, p. 252, Pl. 7:7–9. □1990 *Nuttallides umbonifera* (Cushman) – Thomas *et al.*, Pl. 9:13–16.

*Description.* – Test free, biconvex with acute periphery and narrow keel. Chambers distinct, 6–9 in the last whorl, slightly increasing in size as added. Sutures broad and distinct, slightly curved backward. Wall calcareous, rough, finely perforate. Aperture interiomarginal, starting almost at umbilicus and extending nearly to the peripheral keel.

*Remarks.* – The genetic placement of this species is unclear. Some workers refer this species to *Nuttallides* and others to *Epistominella*. In this study, I have followed Todd (1965) and placed this species in *Nuttallides* rather than in *Epistominella* because of the position of the aperture and the presence of an umbilical plug.

*Ecology.* – *Nuttallides umbonifera* has previously been taken to be an indicator of Antarctic Bottom Water (AABW) in the Atlantic (Lohmann 1978), Pacific, and Indian oceans (Corliss 1979). *Nuttallides umbonifera* is reported from the lower bathyal and abyssal zones in the Atlantic Ocean, whereas in the Pacific and Indian oceans it dominates the fauna in areas of the coldest AABW (Hermelin 1989). It has not been found in the Antarctic or marginal seas, except at abyssal depth (Schnitker 1980; Douglas & Woodruff 1981). Bremer & Lohmann (1982) and Mackensen *et al.* (1990) reported this species in relation to corrosiveness of the bottom waters. *Nuttallides umbonifera,* together with *E. exigua,* are two of the most important deep- and bottom-water species of the Pacific Ocean (Culp 1977; Walch 1978; Schnitker 1980; Hermelin 1989), whereas it occurs in low frequencies in both the South China and Sulu Seas (Miao & Thunell 1993).

Recent studies by Gooday (1993) suggest that it might be informative to compare the distribution of *N. umbonifera* with that of *E. exigua,* a species which is largely controlled by the presence of organic material on the sea floor. *Nuttallides umbonifera* is usually more abundant in areas with a minimal phytodetrial input (Gooday 1993), for example in Northeast Atlantic south of 40°N (Weston & Murray 1994), but it also occurs as far north as 55–57°N (Lukashina 1988). At this latitude in the Northeast Atlantic, the abundances of *N. umbonifera* are less than 5% where *E. exigua* is abundant but reach 31–55% in the Northwest Atlantic, where *E. exigua* is less common (Gooday 1993). *Nuttallides umbonifera* and *E. exigua* also co-occur in the Weddell Sea and South Atlantic, where *N. umbonifera* is dominating below the lysocline and above the CCD and *E. exigua* above the lysocline (Mackensen *et al.* 1990, 1993). Futhermore, *N. umbonifera* is also found to have a diet similar to *E. exigua* (Gooday 1993).

Therefore, it may be suggested that the main ecological difference between these two species is that *N. umbonifera* is a non-opportunist, able to survive, when necessary, on a lower food supply than *E. exigua* (Gooday 1993). Also, *N. umbonifera* has a relatively large, thick-walled test (Mackensen *et al.* 1990), which suggests a slow rate of growth compared with the small, thin-walled tests of *E. exigua.* Thus, *N. umbonifera* may be outcompeted, or at least numerically swamped, by fast-growing opportunists such as *E. exigua* (Gooday 1993). *Nuttallides umbonifera* may, therefore, have greater chance to grow vigorously where food supply is reduced (Loubere 1991) or in carbonate-undersaturated bottom water (Bremer & Lohmann 1982; Mackensen *et al.* 1990).

*Distribution at Sites 502A and 503B.* – *Nuttallides umbonifera* is the most abundant species at both sites. It occurs in most samples, absolute abundances being 0–266 individuals per sample at Site 502A and 0–58 at Site 503B.

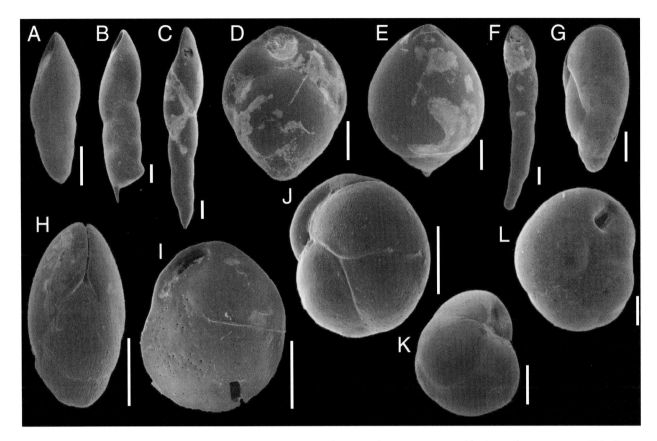

*Fig. 22.* Scale bar 100 μm. □A, B. *Pleurostomella acuminata* Cushman; A, Side view, Hole 502A, 98.81 m; B, Side view, Hole 502A, 98.81 m. □C. *Pleurostomella alternans* Schwager, Side view, Hole 503B, 52.47 m. □D. *Pleurostomella brevis* Schwager, Side view, Hole 502A, 100.73 m. □E. *Pleurostomella recens* Dervieux, Side view, Hole 502A, 98.81 m. □F. *Pleurostomella subnodosa* (Reuss), Side view, Hole 502A, 72.39 m. □G, H. *Francesita advena* (Cushman); G, Side view, Hole 502A, 70.42 m; H, Edge view, Hole 502A, 70.42 m. □I. *Cassidulina carinata* Silvestri, Side view, Hole 502A, 48.80 m. □J. *Cassidulina crassa* d'Orbigny, Side view, Hole 502A, 41.62 m. □K, L. *Globocassidulina subglobosa* (Brady); K, Side view, Hole 503B, 38.13 m; L, Side view, Hole 503B, 38.13 m.

# Superfamily Cassidulinacea d'Orbigny, 1839

## Family Pleurostomellidae Reuss, 1860

## Subfamily Pleurostimellinae Reuss, 1860

## Genus *Pleurostomella* Reuss, 1860

### *Pleurostomella acuminata* Cushman, 1922

Fig. 22A–B

*Synonymy.* – □1922 *Pleurostomella acuminata* n.sp. – Cushman, pp. 50–51, Pl. 19:6. □1953 *Pleurostomella acuminata* Cushman, 1922 – Phleger *et al.*, p. 39, Pl. 8:10–13. □1960 *Pleurostomella acuminata* Cushman, 1922 – Barker, p. 106, Pl. 51:22. □1978 *Pleurostomella acuminata* Cushman, 1922 – Boltovskoy, Pl. 5:39–41. □1989 *Pleuros-* *tomella acuminata* Cushman, 1922 – Hermelin, pp. 70–71, Pl. 13:1.

*Description.* – Test free, fusiform, elongate, biserial, usually widest below the middle. Sutures distinct, slightly depressed. Wall calcareous, smooth, finely perforate. Aperture terminal, almost vertical at the inner face of the ultimate chamber, with a projecting hood.

*Distribution at Sites 502A and 503B.* – *Pleurostomella acuminata* is represented in most of the samples from both sites, but usually in low abundances.

### *Pleurostomella alternans* Schwager, 1866

Fig. 22C

*Synonymy.* – □1866 *Pleurostomella alternans* n.sp. – Schwager, p. 238, Pl. 6:79. □1927 *Pleurostomella alternans* Schwager, 1866 – Cushman & Harris, p. 129, Pl. 25:7–8. □1964 *Pleurostomella alternans* Schwager, 1866 – Leroy, p. 36, Pl. 5:5. □1966 *Pleurostomella alternans* Schwager,

1866 – Todd, p. 29, Pl. 12:14–15. □1978 *Pleurostomella alternans* Schwager, 1866 – Boltovskoy, Pl. 5:43–44. □1986 *Pleurostomella alternans* Schwager, 1866 – Boersma, Pl. 5:6. □1989 *Pleurostomella alternans* Schwager, 1866 – Hermelin, p. 71.

*Description.* – Test free, biserial, elongate, widest at the apertural end, tapering towards the initial end. Sutures distinct, slightly depressed. Wall calcareous, smooth, finely perforate. Aperture terminal, at the inner face of the ultimate chamber, with a projecting hood.

*Remarks.* – *Pleurostomella alternans* has a more tapered test than *P. acuminata,* with the widest part at the apertural end.

*Distribution at Sites 502A and 503B.* – *Pleurostomella alternans* is a rare species with scattered occurrences at both Site 502A and Site 503B.

## *Pleurostomella brevis* Schwager, 1866

Fig. 22D

*Synonymy.* – □1866 *Pleurostomella brevis* n.sp. – Schwager, p. 239, Pl. 6:81. □1884 *Pleurostomella brevis* Schwager, 1866 – Brady, p. 411, Pl. 51:20a–b. □1960 *Pleurostomella brevis* Schwager, 1866 – Barker, p. 104, Pl. 51:20a–b. □1964 *Pleurostomella brevis* Schwager, 1866 – Leroy, p. 36, Pl. 5:4. □1986 *Pleurostomella brevis* Schwager, 1866 – Belanger & Berggren, p. 340. □1989 *Pleurostomella brevis* Schwager, 1866 – Hermelin, p. 71.

*Description.* – Test free, subcylindrical, broadest at the middle of test, ovate in cross-section. Chambers strongly embracing, the last chamber occupying more than half the length of the test. Sutures slightly depressed. Wall calcareous, hyaline, smooth, finely perforate. Aperture a vertical slit in the depressed face of the ultimate chamber.

*Distribution at Sites 502A and 503B.* – *Pleurostomella brevis* is represented by a single specimen at Site 502A (100.73 m). It is rare with scattered occurrences at Site 503B.

## *Pleurostomella recens* Dervieux, 1899

Fig. 22E

*Synonymy.* – □1884 *Pleurostomella rapa* Guembel – Brady, p. 411, Pl. 51:21a–b. □1899 *Pleurostomella rapa* Guembel var. *recens* Dervieux, p. 76. □1904b *Ellipsopleurostomella pleurostomella* Silvestri, p. 8, Figs. 4–5. □1927c *Pleurostomella rapa* Guembel var. *recens* – Cushman p. 156, Pl. 28:6a–b. □1927 *Pleurostomella pleurostomella* Silvestri – Cushman & Harris, p. 130, Pl. 25:13. □1936 *Pleurostomella bierigi* n.sp. – Palmer & Bermúdez, p. 294, Pl.

17:7–8. □1960 *Pleurostomella rapa* Guembel var. *recens* – Barker, p. 104, Pl. 51:21a–b. □1978 *Pleurostomella bierigi* Palmer & Bermúdez, 1936 – Boltovskoy, Pl. 5:45. □1989 *Pleurostomella recens* Dervieux – Hermelin, pp. 71–72, Pl. 13:2.

*Description.* – Test free, small. Broadest in the middle of test, apical end pointed. Chambers strongly embracing, the ultimate chamber occupying most of the test. Sutures slightly depressed. Wall calcareous, hyaline, smooth, finely perforate. Aperture a vertical slit in the depressed face of the ultimate chamber.

*Remarks.* – The apical end is more pointed and the chambers are more embracing in *Pleurostomella recens* than in *P. brevis*. According to Hermelin (1989), *P. bierigi* is regarded as a junior synonym of *P. recens.*

*Distribution at Sites 502A and 503B.* – *Pleurostomella recens* is rare with scattered occurrences at both investigated sites.

## *Pleurostomella subnodosa* (Reuss, 1845)

Fig. 22F

*Synonymy.* – □1845 *Nodosaria nodosa* – Reuss (part), p. 28, Pl. 13:22. □1851b (part) *Dentalina subnodosa* (Reuss) – Reuss, p. 24, Pl. 1:9. □1860 *Pleurostomella subnodosa* (Reuss) – Reuss, p. 204, Pl. 8:2a–b. □1884 *Pleurostomella subnodosa* (Reuss) – Brady, pp. 412–413, Pl. 52:12–13. □1911 *Pleurostomella subnodosa* (Reuss) – Cushman, p. 51, Text-fig. 82. □1960 *Pleurostomella* sp. – Barker, Pl. 52:12–13. □1989 *Pleurostomella subnodosa* (Reuss) – Hermelin, p. 72, Pl. 13:7.

*Description.* – Test free, elongated, early biserial, later uniserial. Initial portion rounded, aperture portion subacute in front view, circular in side view. Sutures slightly depressed. Wall calcareous, hyaline, smooth, finely perforate. Aperture terminal, sinus broad with slight projections at each side.

*Distribution at Sites 502A and 503B.* – At Site 502A *Pleurostomella subnodosa* is most abundant in the interval between 119.33 and 122.32 m; in other intervals it is rare with scattered occurrences. At Site 503B it is rare with few occurrences.

## Family Caucasinidae Bykova, 1959

## Subfamily Causasininae Bykova, 1959

## Genus *Francesita* Loeblich & Tappan, 1963

### *Francesita advena* (Cushman, 1922)

Fig. 22G–H

*Synonymy.* – □1922 *Virgulina* (?) *advena* n.sp. – Cushman, p. 120, Pl. 25:1–3. □1937c *Virgulina* (?) *advena* Cushman, 1922 – Cushman, p. 29, Pl. 4:29. □1953 *Francesita advena* (Cushman) – Loeblich & Tappan, p. 215. □1953 *Virgulina advena* Cushman, 1922 – Phleger *et al.*, p. 34, Pl. 7:1–2. □1976 *Francesita advena* (Cushman) – Pflum & Frerichs, Pl. 4:6–7. □1985 *Francesita advena* (Cushman)– Thomas, p. 676, Pl. 2:8. □1989 *Francesita advena* (Cushman)– Hermelin, p. 72, Pl. 13:8–10.

*Description.* – Test free, elongate, body subcylindrical with broadly rounded base, circular to oval in cross-section. Triserial in early stage, later biserial. Sutures slightly depressed to flush. Wall calcareous, surface smooth, finely perforate. Aperture an elongate slit extending from the base of the final chamber up across the top, about half-way down the opposite side, with one margin of the aperture being curved inward and the opposite one projecting above like a narrow hood.

*Ecology.* – According to Pflum & Frerichs (1976), *Francesita advena* is the deepest-water 'virgulinid' index known, and it has been found to be almost exclusively an abyssal species in the Gulf of Mexico.

*Distribution at Sites 502A and 503B.* – *Francesita advena* is a relatively common species with absolute abundances of 0–6 individuals in each sample at Site 502A. It is rare with few occurrences at Site 503B.

## Family Cassidulinidae d'Orbigny, 1839

## Genus *Cassidulina* d'Orbigny, 1826

### *Cassidulina carinata* Silvestri, 1896

Fig. 22I

*Synonymy.* – □1884 *Cassidulina laevigata* d'Orbigny – Brady, pp. 428–429, Pl. 54:2–3. □1896 *Cassidulina laevigata* var. *carinata* n.sp. – Silvestri, p. 104, Pl. 2:10a–c. □1923 *Cassidulina laevigata carinata* n.sp. – Cushman, p. 124, Pl. 25:6–7. □1950 *Cassidulina neocarinata* sp.nov. – Thalmann, p. 44. □1950 *Cassidulina laevigata* d'Orbigny – Parr, p. 343. □1951 *Cassidulina laevigata carinata* – Phleger & Parker, p. 27, Pl. 14:7a–b. □1953 *Cassidulina carinata* Silvestri – Phleger *et al.*, p. 44, Pl. 9:32, 37. □1962 *Cassidulina laevigata* d'Orbigny – McKnight, p. 127, Pl. 21:140a–b (not *Cassidulina laevigata* d'Orbigny 1826). □1965 *Cassidulina carinata* Silvestri – Todd, pp. 40–41, Pl. 17:4. □1980 *Cassidulina carinata* Silvestri – Bremer *et al.*, pp. 23–24, Pl. 2:9–10. □1980 *Cassidulina laevigata carinata* Silvestri – Ingle *et al.*, p. 131, Pl. 6:5–8. □1985 *Cassidulina neocarinata* Thalmann, 1950 – Kohl, p. 86, Pl. 30:1. □1985 *Cassidulina carinata* Silvestri – Mead, p. 232, Pl. 3:1a–3b.

*Description.* – Test free, compressed, small; chambers biserially arranged and enrolled, tightly coiled. Chambers elongate, about four pairs of chambers in the final whorl, slightly inflated; periphery slightly lobulate in side view, acute with a narrow keel. Sutures distinct, slightly depressed. Wall calcareous, hyaline, smooth, coarsely perforate. Aperture an elongate broad slit parallel to the periphery along the suture of the last chamber, with a tooth.

*Remarks.* – *Cassidulina carinata* Silvestri has a broader apertural face and tooth and a less compressed test than *Cassidulina neocarinata* Thalmann and is more coarsely perforate (Phleger *et al.* 1953). Phleger *et al.* (1953) also pointed out that neither the aperture nor the apertural face of *C. carinata* resembles that of *C. laevigata* d'Orbigny, the type specimen of which has a more rounded apertural face and a small aperture. Mead (1985) was of the opinion that the type specimen of *C. laevigata* appears to have been a specimen with a broken final chamber, so its apertural characteristics cannot be ascertained. Todd (1965) concluded that these species overlap in morphologic characteristics and that *C. carinata* and *C. neocarinata* should be considered as synonyms. Eade (1967) agreed with this opinion and stated that intraspecific variation includes the characteristics used to separate these two species. Rodrigues *et al.* (1980, Pl. 5:2, 5, 8) illustrated a specimen of *C. neocarinata*, which is distinctly different from *C. carinata* (their Pl. 5:1, 4, 7), and stated that *C. neocarinata* does not have the broad flap characteristic of *C. carinata*. However, the specimens from the present study closely match *C. carinata* as described by Phleger *et al.* (1953). They are generally small, narrow, with an apertural tooth-plate, a narrow, nonpunctate keel, and about 3.5–4 pairs of chambers per whorl.

*Ecology.* – The bathymetric distribution of this species is primarily between 500 and 2500 m. It is cosmopolitan but usually occurs in low abundances (Mead 1985). In the South China Sea, *Cassidulina* dominates the fauna at water depths between 1,700 and 2,800 m, where the organic-carbon content of the sediment is high because of higher input of terrestrial organic carbon and higher sedimentation rates (Miao & Thunell 1993). Also, *Cassidulina* is more abundant during glacial stage 2 than during the Holocene in the South China Sea (Miao & Thunell 1996).

*Distribution at Sites 502A and 503B. – Cassidulina carinata* is a rare species with few occurrences at Site 502A. It is absent at Site 503B.

## *Cassidulina crassa* d'Orbigny, 1839

Fig. 22J

*Synonymy.* – ☐1839b *Cassidulina crassa* n.sp. – d'Orbigny, p. 56, Pl. 7:18–20. ☐1884 *Cassidulina crassa* d'Orbigny, 1839b – Brady, p. 429, Pl. 54:5. ☐1953 *Cassidulina crassa* d'Orbigny, 1839b – Phleger *et al.*, pp. 44–45, Pl. 10:1. ☐1958 *Cassidulina crassa* d'Orbigny, 1839b – Nørvang, Pl. 8–9:20–25. ☐1978 *Cassidulina crassa* d'Orbigny, 1839b – Wright, p. 712, Pl. 3:11–12. ☐1986 *Globocassidulina* cf. *G. crassa* (d'Orbigny) – Belanger & Berggren, p. 340.

*Description.* – Test closely coiled, lenticular or subglobular, only slightly compressed; periphery slightly lobate. Chambers slightly depressed. Sutures distinct, slightly curved, depressed. Wall calcareous, hyaline, smooth, finely perforate. Aperture an elongate narrow slit parallel to the periphery along the suture of the last chamber.

*Remarks. – Cassidulina crassa* differs from *C. minuta* because it has a less compressed test. Also, the aperture of *C. minuta* is more elongate than that of *C. crassa*.

*Distribution at Sites 502A and 503B. – Cassidulina crassa* is a very rare species with a few occurrences above 68.42 m at Site 502A. It is absent at Site 503B.

## Genus *Globocassidulina* Voloshinova, 1960

## *Globocassidulina subglobosa* (Brady, 1881)

Fig. 22K–L

*Synonymy.* – ☐1881 *Cassidulina subglobosa* n.sp. – Brady, p. 60. ☐1884 *Cassidulina subglobosa* Brady, 1881 – Brady, p. 430, Pl. 54:17a–c. ☐1911 *Cassidulina subglobosa* Brady, 1881 – Cushman, p. 98, Text-fig. 152. ☐1921 *Cassidulina subglobosa* Brady, 1881 – Cushman, pp. 171–172, Pl. 32:2. ☐1925b *Cassidulina subglobosa* Brady, 1881 – Cushman, p. 54, Pl. 8:48–50. ☐1941 *Cassidulina subglobosa* Brady, 1881 – Galloway & Heminway, p. 425, Pl. 32:2a–b. ☐1945

*Cassidulina subglobosa* Brady, 1881 – Cushman & Todd, p. 61, Pl. 10:8a–b. ☐1951 *Cassidulina subglobosa* Brady, 1881 – Phleger & Parker, p. 27, Pl. 14:11–12. ☐1953 *Cassidulina subglobosa* Brady, 1881 – Phleger *et al.*, p. 45, Pl. 10:4. ☐1964 *Islandiella subglobosa* (Brady) Akers & Dorman, p. 39, Pl. 11:19. ☐1966 *Globocassidulina subglobosa* (Brady) – Belford, p. 149, Pl. 25:11–16, Text-figs. 17:1–7; 18:1–4. ☐1971 *Islandiella subglobosa* (Brady) – Schnitker, p. 204, Pl. 5:1a–c. ☐1978 *Cassidulina subglobosa subglobosa* Brady – Boltovskoy, p. 155, Pl. 2:34. ☐1978 *Globocassidulina subglobosa* (Brady) – Lohmann, p. 26, Pl. 2:8–9. ☐1979 *Globocassidulina subglobosa* (Brady) – Corliss, p. 8, Pl. 3:12–13. ☐1980 Cassidulina subglobosa Brady, 1881 – Butt, Pl. 2:11. ☐1980 *Cassidulina subglobosa subglobosa* Brady – Ingle *et al.*, p. 132, Pl. 1:13–14. ☐1981 Cassidulina subglobosa Brady, 1881 – Resig, Pl. 7:7. ☐1985 *Globocassidulina subglobosa* (Brady) – Kohl, p. 88, Pl. 30:3–4. ☐1985 *Globocassidulina subglobosa* (Brady) – Mead, p. 232, 234, Pl. 3:8. ☐1985 *Globocassidulina subglobosa* (Brady) – Thomas, Pl. 7:4. ☐1986 *Globocassidulina subglobosa* (Brady) – Kurihara & Kennett, Pl. 5:4–8. ☐1989 *Globocassidulina subglobosa* (Brady) – Hermelin, pp. 74–75. ☐1990 *Globocassidulina subglobosa* (Brady) – Ujiié, pp. 39–40, Pl. 21:4, 5a–b; 6a–b; 7a–b; Pl. 22:1a–b.

*Description.* – Test free, subglobular in side view, coiled biserially, ovate in transverse section. Chambers inflated, 4–5 pairs in last whorl, increasing gradually in size as added. Sutures narrow, distinct, smooth, slightly curved. Wall calcareous, hyaline, smooth, finely perforate. Aperture narrow, in a depression of the apertural face, varying in shape from a straight, narrow slit to a chevron-shaped opening, with a lip attached to the outer margin; toothplate formed by infolding of apertural face.

*Ecology. – Globocassidulina subglobosa* is a cosmopolitan species present over a wide bathymetric range and in a number of different water masses. It has been stated to be an indicator for NADW in the North Atlantic (Streeter 1973; Lohmann 1978; Schnitker 1979; Hermelin 1986). It has also been found to be a dominant form in the MOW (Mediterranean Outflow Water) (Murray 1991) and was considered a typical species of 'warm AABW' in the Indian Ocean (Corliss 1979). In the Gulf of Mexico, *G. subglobosa* occurs in the upper bathyal zone (Pflum & Frerichs 1976). According to Corliss (1985, 1991) and Corliss & Chen (1988), *G. subglobosa* may be characterized as a shallow-infaunal species. It is considered to thrive in sediments with high organic-carbon contents and can tolerate low dissolved-oxygen concentration (Ingle *et al.* 1980; Miller & Lohmann 1982; Corliss & Chen 1988). In the South China Sea and Sulu Sea, *G. subglobosa* is found within the oxygen minimum zone of each basin where the organic-carbon content of the sediment is highest and the oxygen penetration depth is shallowest (Miao & Thunell 1993). However, Loubere & Banonis (1987) and Burke *et*

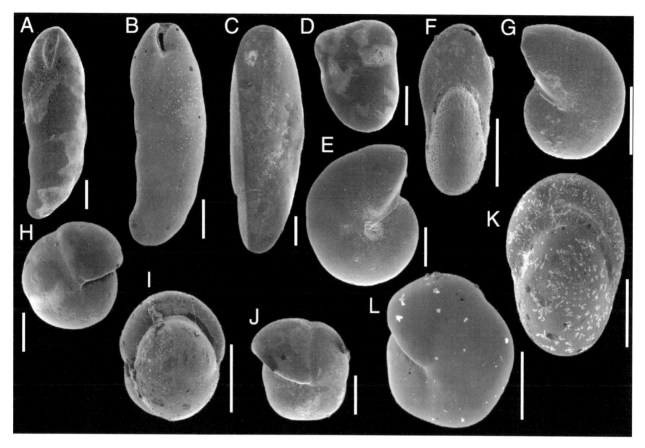

*Fig. 23.* Scale bar 100 μm. □A. *Rutherfordoides bradyi* (Cushman), Side view, Hole 502A, 91.85 m. □B. *Rutherfordoides tenuis* (Phleger & Parker), Side view, Hole 502A, 109.51 m. □C. *Chilostomella oolina* Schwager, Side view, Hole 502A, 100.73 m. □D. *Allomorphina pacifica* Cushman & Todd, Ventral view, Hole 503B, 72.03 m. □E–G. *Nonion germanicum* (Ehrenberg); E, Side view, Hole 502A, 76.72 m; F, Edge view, Hole 502A, 76.72 m; G, Opposite side view, Hole 502A, 76.72 m. □H–J. *Pullenia bulloides* (d'Orbigny); H, Side view, Hole 502A, 65.51 m; I, Edge view, Hole 502A, 71.42 m; J Opposite side view, Hole 502A, 65.51 m. □K, L. *Pullenia quinqueloculina* (Reuss); K, Edge view, Hole 502A, 110.47 m; L, Side view, Hole 502A, 110.47 m.

al. (1993) reported from studies in the eastern and western equatorial Pacific that *G. subglobosa* may be related to environment with low productivity in the surface area and thereby low flux of organic matter to the sea-floor. Miao & Thunell (1996) found that *G. subglobosa* is absent at the end of glacial stage 2 and is present in low abundances (<5%) during the Holocene in the Sulu Sea, whereas it is abundant only in the late Holocene in the South China Sea.

*Distribution at Sites 502A and 503B.* – *Globocassidulina subglobosa* is a relatively common species at both sites. Absolute abundances 0–33 individuals per sample at Site 502A and 0–13 at Site 503B.

## Genus *Rutherfordoides* McCulloch, 1981

## *Rutherfordoides bradyi* (Cushman, 1922)

Fig. 23A

*Synonymy.* – □1922 *Virgulina bradyi* n.sp. – Cushman, p. 115, Pl. 24:1. □1937c *Virgulina bradyi* Cushman, 1922 – Cushman, p. 29, Pl. 5:1a–c. □1953 *Virgulina bradyi* Cushman, 1922 – Phleger *et al.*, p. 34, Pl. 7:4–5. □1960 *Virgulina bradyi* Cushman, 1922 – Barker, p. 106, Pl. 52:9a–c. □1989 *Rutherfordoides bradyi* (Cushman) – Hermelin, pp. 75–76, Pl. 14:7–8.

*Description.* – Test free, slender, circular to ovate in transverse section, early portion coiled; chambers biserially enrolled, later portion uncoiled. Chambers distinct, slightly inflated, increasing gradually in size. Sutures distinct, flush with surface. Wall calcareous, hyaline, smooth, perforated by small, narrow pores. Aperture a narrow loop in a depression along the suture, extending from base of ultimate chamber to apex of test.

*Remarks.* – *Rutherfordoides bradyi* is similar to *R. tenuis,* but the test is more slender.

*Distribution at Sites 502A and 503B.* – *Rutherfordoides bradyi* is a very rare species with few occurrences at both investigated sites.

## *Rutherfordoides tenuis* (Phleger & Parker, 1951)

Fig. 23B

*Synonymy.* – □1951 *Cassidulinoides tenuis* n.sp. – Phleger & Parker, p. 27, Pl. 14:14–17. □1958 *Cassidulinoides tenuis* Phleger & Parker, 1951 – Parker, p. 272, Pl. 4:18–19. □1963 *Cassidulinoides tenuis* Phleger & Parker, 1951 – Matsunaga, Pl. 49:6a–b. □1964 *Cassidulinoides tenuis* Phleger & Parker, 1951 – Leroy, p. 41, Pl. 12:1–2. □1978 *Cassidulinoides tenuis* Phleger & Parker, 1951 – Boltovskoy, Pl. 3:2. □1985 *Rutherfordoides tenuis* (Phleger & Parker) – Kohl, pp. 89–90, Pl. 18:5a–f. □1989 *Rutherfordoides tenuis* (Phleger & Parker) – Hermelin, p. 75, Pl. 14:3–4.

*Description.* – Test free, elongate; length 2.5 times the width, early portion coiled; chambers biserially enrolled, later portion uncoiled, circular to ovate in transverse section. Chambers distinct, slightly inflated, increasing gradually in size. Sutures distinct, slightly depressed. Wall calcareous, hyaline, smooth, perforated with small, narrow pores. Aperture an elongate opening in a depression, extending from base of ultimate chamber to a point near the apex of test.

*Remarks.* – *Rutherfordoides tenuis* differs from *R. bradyi* in its rounded cross-section and more elongate shape.

*Ecology.* – *Rutherfordoides tenuis* has been found in the upper bathyal zone off western Central America (Smith 1964).

*Distribution at Sites 502A and 503B.* – *Rutherfordoides tenuis* is a very rare species with few occurrences at both Site 502 and Site 503.

# Superfamily Nonionacea Schultze, 1854

# Family Nonionidae Schultze, 1854

# Subfamily Chilostomellinae Brady, 1881

# Genus *Chilostomella* Reuss *in* Czjzek, 1849

## *Chilostomella oolina* Schwager, 1878

Fig. 23C

*Synonymy.* – □1878 *Chilostomella oolina* n.sp. – Schwager, p. 527, Pl. 1:16. □1926a *Chilostomella oolina* Schwager, 1878 – Cushman, p. 74, Pl. 11:3–10. □1953 *Chilostomella oolina* Schwager, 1878 – Phleger *et al.*, p. 47, Pl. 10:18. □1960 *Chilostomella oolina* Schwager, 1878 – Barker, p. 112, Pl. 55:12–14, 17–18. □1971 *Chilostomella oolina* Schwager, 1878 – Schnitker, p. 196, Pl. 10:3a–b. □1980 *Chilostomella oolina* Schwager, 1878 – Ingle *et al.*, p. 132, Pl. 6:9–10. □1986 *Chilostomella oolina* Schwager, 1878 – Kurihara & Kennett, Pl. 6:10. □1989 *Chilostomella oolina* Schwager, 1878 – Hermelin, p. 76, Pl. 14:5. □1990 *Chilostomella oolina* Schwager, 1878 – Thomas *et al.*, Pl. 7:9. □1990 *Chilostomella oolina* Schwager, 1878 – Ujiié, p. 41, Pl. 22:5–6.

*Description.* – Test free, elongate, about three times as long as broad, both ends broadly rounded, sides nearly parallel. Ultimate chamber comprises more than half of the test. Wall calcareous, hyaline, thin, translucent, finely perforated. Aperture a narrow, curved slit at the suture between the ultimate and penultimate chambers.

*Ecology.* – According to Pflum & Frerichs (1976), the size of *C. oolina* increases with increasing water depth, from about 0.4 mm near its upper depth limit in the lower neritic zone to approximately 0.6 mm in the lower middle and lower bathyal and abyssal zones in the Gulf of Mexico. Previous studies have classified *Chilostomella* as an infaunal taxon (Corliss 1985; Corliss & Fois 1990) adapted to low oxygen conditions (e.g., Corliss 1985; Mackensen & Douglas 1989). According to Corliss (1991), *C. oolina* is a deep-infaunal species, extending to sediment depths below 4 cm in the North Atlantic. It can also live at shallower depths in the sediment depending on environmental conditions (Rathburn & Corliss 1994). Rathburn & Corliss (1994) suggested that *Chilostomella* is adapted to take advantage of subsurface accumulations of labile organic carbon, even in deep turbiditic sediments. Miao & Thunell (in press) found that *C. oolina* has its highest abundance in the middle Holocene (17%) in the South China Sea.

*Distribution at Sites 502A and 503B. – Chilostomella oolina* is a rare species with scattered occurrences at both Site 502A and Site 503B.

# Genus *Allomorphina* Reuss, *in* Czjzek, 1849

## *Allomorphina pacifica* Cushman & Todd, 1949

Fig. 23D

*Synonymy. –* □1884 *Allomorphina trigona* (Reuss) – Brady, p. 438, Pl. 55:24–26. □1914 *Allomorphina trigona* (Reuss) – Cushman, p. 3, Pl. 1:6–8. □1925a *Allomorphina trigona* (Reuss) – Cushman, p. 133, Pl. 17:2. □1944 *Valvulineria* aff. *allomorphinoides* (Reuss) – Leroy, p. 87, Pl. 3:21–23. □1949 *Allomorphina pacifica* sp.nov. – Cushman & Todd, pp. 68–69, Pl. 12:6–9. □1951 *Allomorphina pacifica* nom.nov. – Hofker, p. 138:86a–f. □1960 *Allomorphina pacifica* Cushman & Todd, 1951 – Barker, p. 112, Pl. 55:24–26. □1978 *Allomorphina pacifica* Cushman & Todd, 1951 – Boltovskoy, Pl. 1:1. □1980 *Valvulineria* sp. Butt, Pl. 6:19–20. □1986 *Allomorphina pacifica* Cushman & Todd, 1951 – Kurihara & Kennett, Pl. 6:9. □1989 *Allomorphina pacifica* Cushman & Todd, 1951 – Hermelin, p. 76, Pl. 14:1–2. □1990 *Allomorphina pacifica* Cushman & Todd, 1951 – Ujiié, p. 41, Pl. 22:7–11.

*Description. –* Test free, trochospiral, involute, ovate in side view and in transverse section. Sutures distinct, slightly depressed. Wall calcareous, hyaline, smooth. Aperture a low, narrow opening at the sides of a V-shaped lip, edge of lip saw-toothed.

*Remarks. –* According to Hermelin (1989), *A. pacifica* is probably restricted to the Pacific Ocean.

*Distribution at Sites 502A and 503B. – Allomorphina pacifica* is represented by only one specimen at Site 503B (72.03 m).

# Subfamily Nonoininae Schultze, 1854

# Genus *Nonion* de Montfort, 1808

## *Nonion germanicum* (Erhenberg, 1840)

Fig. 23E–G

*Synonymy. –* □1840 *Nonionina germanica* n.sp. – Ehrenberg, p. 23. □1953 *Nonion germanicum* (Erhenberg) – Phleger *et al.*, p. 30, Pl. 6:6. □1979 *Nonion germanicum* (Erhenberg) – Corliss, p. 8, Pl. 3:14–15.

*Description. –* Test free, planispiral, involute, periphery rounded. Chambers distinct, usually about eight in the last whorl, increasing rapidly in size. Sutures gently curved, flush with surface. Wall calcareous, hyaline, smooth; the umbilicus is closed and surrounded by clear calcite. The periphery of the early chambers is more acute than with the later chambers, and the chambers also rapidly increase in size. Aperture an interiomarginal slit that extends to umbilicus on both sides.

*Ecology. – Nonion germanicum* is generally considered a bathyal species. In the Indian Ocean it is found in the bathyal zone and down into the abyssal zone (Corliss, 1979).

*Distribution at Sites 502A and 503B. – Nonion germanicum* is a relatively common species at Site 502A. In contrast, it has more scattered occurrences at Site 503B.

# Genus *Astrononion* Cushman & Edwards, 1937

## *Astrononion gallowayi* Loeblich & Tappan, 1953

*Synonymy. –* □1937 *Astrononion australe* n.sp. – Cushman & Edwards, pp. 33–34, Pl. 3:13–14. □1939a *Astrononion australe* Cushman & Edwards, 1937 – Cushman, pp. 37–38, Pl. 10:7–8. □1937 *Astrononion sidebottomi* n.sp. – Cushman & Edwards, pp. 31–32, Pl. 3:8. □1937 *Astrononion stellatum* n.sp. – Cushman & Edwards, p. 32, Pl. 3:9–11. □1937 *Astrononion tumidum* n.sp. – Cushman & Edwards, p. 33, Pl. 3:17. □1937 *Astrononion viragoense* n.sp. – Cushman & Edwards, pp. 32–33, Pl. 3:12. □1939a *Astrononion sidebottomi* Cushman & Edwards, 1937 – Cushman, p. 36, Pl. 10:2. □1939a *Astrononion stellatum* Cushman & Edwards, 1937 – Cushman, p. 36, Pl. 10:3–5. □1939a *Astrononion tumidum* Cushman & Edwards, 1937 – Cushman, p. 37, Pl. 10:11. □1939a *Astrononion viragoense* Cushman & Edwards, 1937 – Cushman, p. 36, Pl. 10:6. □1953 *Astrononion gallowayi* nom.nov. – Loeblich & Tappan, p. 90, Pl. 17:4–7. □1964 *Astrononion gallowayi* Loeblich & Tappan, 1953 – Feyling–Hanssen, p. 332, Pl. 18:4. □1971 *Astrononion stellatum* Cushman & Edwards, 1937 – Schnitker, p. 193, Pl. 10:7a–b (not *Nonionina stellata* Terquem, 1882). □1981 *Astrononion gallowayi* Loeblich & Tappan, 1953 – Cole, p. 109, Pl. 13:6. □1985 *Astrononion gallowayi* Loeblich & Tappan, 1953 – Hermelin & Scott, p. 203, Pl. 5:1a–c. □1985 *Astrononion gallowayi* Loeblich & Tappan, 1953 – Kohl, p. 91, Pl. 32:1.

□1989 *Astrononion gallowayi* Loeblich & Tappan, 1953 – Hermelin, p. 77. □1990 *Astrononion gallowayi* Loeblich & Tappan, 1953 – Thomas *et al.*, p. 226.

*Description.* – Test free, planispiral and involute, compressed, umbilical area concave, periphery rounded. Chambers distinct, 9–10 in final whorl, increasing gradually in size, strongly inflated. Sutures depressed. Wall calcareous, smooth, finely perforated. Aperture a low arch at the base of the final chamber, supplementary aperture at outer posterior margin of each supplementary chamber.

*Distribution at Sites 502A and 503B.* – *Astrononion gallowayi* is represented by a single specimen at Site 503B (38.13 m).

## Genus *Florilus* Montfort, 1808

## *Florilus atlanticus* (Cushman, 1947)

*Synonymy.* – □1947 *Nonionella atlantica* n.sp. – Cushman, p. 90, Pl. 20:4–5. □1951 *Nonionella atlantica* Cushman, 1947 – Phleger & Parker (part), p. 11, Pl. 5:21a–b, 22a–b (not Pl. 5:23a–b). □1953 *Nonionella atlantica* Cushman, 1947 – Phleger *et al.*, p. 31, Pl. 6:9–10. □1954 *Nonionella atlantica* Cushman, 1947 – Parker, p. 507, Pl. 6:6–7. □1957 *Nonionella atlantica* Cushman, 1947 – Todd & Brönnimann, p. 32, Pl. 5:30–31. □1961 *Pseudononion atlantica* (Cushman) – Andersen, p. 84, Pl. 18:1a–b, 2a–c. □1964 *Florilus atlanticus* (Cushman) Akers & Dorman, p. 34, Pl. 6:26–27. □1976 *Nonionella atlantica* Cushman, 1947 – Hansen & Lykke–Andersen, p. 23, Pl. 21:9–12. □1985 *Florilus atlanticus* (Cushman)– Kohl, p. 91, Pl. 32:2. □1990 *Nonionella atlantica* Cushman, 1947 – Thomas *et al.*, p. 227.

*Description.* – Test free, compressed, asymmetrical; dorsal side showing earlier coils, which are covered on the ventral side; periphery rounded, umbilical region on ventral side slightly depressed. Chambers distinct, inflated, 9–11 in the last whorl, increasing rapidly in size as added. Sutures distinct, slightly depressed, curved. Wall calcareous, hyaline, smooth, finely perforate. Aperture a narrow, interiomarginal, equatorial opening.

*Distribution at Sites 502A and 503B.* – *Florilus atlanticus is* a rare species with scattered occurrences at Site 502A. It is absent at Site 503B.

## Genus *Pullenia* Parker & Jones, *in* Carpenter, Parker & Jones, 1862

## *Pullenia bulloides* (d'Orbigny, 1846)

Fig. 23H–J

*Synonymy.* – □1846 *Nonionina bulloides* d'Orbigny – d'Orbigny, p. 107, Pl. 5:9–10. □1862 *Pullenia sphaeroides* Parker & Jones, *in* Carpenter – Parker & Jones, p. 184, Pl. 12:12. □1866 *Pullenia bulloides* (d'Orbigny) – Reuss, p. 150. □1884 *Pullenia sphaeroides* (d'Orbigny) – Brady, pp. 615–616, Pl. 84:12–13. □1943 *Pullenia bulloides* (d'Orbigny) – Cushman & Todd, pp. 13–14, Pl. 2:15–18. □1953 *Pullenia bulloides* (d'Orbigny) – Phleger *et al.*, p. 47, Pl. 10:19. □1960 *Pullenia bulloides* (d'Orbigny) – Barker, p. 174, □1964 *Pullenia bulloides* (d'Orbigny) – Akers & Dorman, p. 49, Pl. 11:11–12. □1964 *Pullenia bulloides* (d'Orbigny) – Leroy, p. 41, Pl. 10:30–31. □1964 *Pullenia miocenica* Kleinpell – Leroy, p. 41, Pl. 10:26–27.Pl. 84:12–13. □1973 *Pullenia bulloides* (d'Orbigny) – Douglas, Pl. 8:1–2. □1978 *Pullenia bulloides* (d'Orbigny) – Boltovskoy, Pl. 6:12. □1978 *Pullenia bulloides* (d'Orbigny) – Lohmann, p. 26, Pl. 1:10–11. □1979 *Pullenia bulloides* (d'Orbigny) – Corliss, p. 8, Pl. 4:1–2. □1981 *Pullenia bulloides* (d'Orbigny) – Burke, Pl. 3:5–6. □1981 *Pullenia bulloides* (d'Orbigny) – Cole, p. 111, Pl. 14:5. □1981 *Pullenia bulloides* (d'Orbigny) – Resig, Pl. 7:13. □1984 *Pullenia bulloides* (d'Orbigny) – Murray, Pl. 2:19–20. □1985 *Pullenia bulloides* (d'Orbigny) – Hermelin & Scott, p. 216, Pl. 5:3–4. □1985 *Pullenia bulloides* (d'Orbigny) – Kohl, pp. 92–93, Pl. 32:5. □1985 *Pullenia bulloides* (d'Orbigny) – Mead, 1985, p. 236, Pl. 4:6. □1986 *Pullenia bulloides* (d'Orbigny) – Kurihara & Kennett, Pl. 6:5–6. □1989 *Pullenia bulloides* (d'Orbigny) – Hermelin, pp. 78–79, Pl. 15:4–5. □1990 *Pullenia bulloides* (d'Orbigny) – Thomas *et al.*, Pl. 7:10, Pl. 10:15. □1991 *Pullenia bulloides* (d'Orbigny) – Scott & Vilks, p. 32.

*Description.* – Test free, planispiral, involute, sphaeroidal, circular in side view, ovate in apertural view. Chambers distinct, inflated, 4–5 in the last whorl. Sutures distinct, radial, flush with surface. Wall calcareous, hyaline, smooth, finely perforate. Aperture a narrow interiomarginal slit extending from one umbilicus to the other, with a thin lip.

*Remarks.* – *Pullenia bulloides* is characterized by a spheroid, smooth outline and usually 4–4.5 chambers.

*Ecology.* – *Pullenia bulloides* is the most common pullenid in the deep-sea (Corliss 1979). Smith (1964) and Ingle & Keller (1980) reported this species from the lower middle bathyal zone in the Pacific, and Pflum & Frerichs (1976) reported it from the neritic zone down into the abyssal zone in the Gulf of Mexico. In the southeast Indian Ocean, *P. bulloides* is an important deep-sea foraminifer and is found at 2,500–4,600 m (Corliss 1979). In the Norwegian Sea off southwest Norway, *P. bulloides* has been found abundant at water depths between 1,000 and 1,400 m, where the slope is covered by organic-rich terrigeneous mud (Mackensen *et al.* 1985). In the Sulu Sea, Rathburn & Corliss (1994) found this shallow infaunal species down to 2 cm, with a maximum at 1.5 cm. According to Burke *et al.* (1993), *P. bulloides* may be associated with areas of low productivity in the surface water and, therefore, low flux of organic matter to the sea-floor, whereas Mackensen *et al.* (1985) reported *P. bulloides* to be more abundant in regions with high productivity in the surface water. It is also abundant in postglacial sediments on the Ontong Java Plateau (Burke *et al.* 1993).

*Distribution at Sites 502A and 503B.* – *Pullenia bulloides* is a common species at both sites, showing absolute abundances of 0–11 individuals per sample at Site 502A and 0–10 at Site 503B.

## *Pullenia quinqueloba* (Reuss, 1851a)

Fig. 23K–L

Synonymy.– □1851a *Nonionina quinqueloba* Reuss, p. 71, Pl. 5:31a–b. □1880 *Pullenia compressa* n.sp. – Seguenza, p. 307, Pl. 17:14. □1943 *Pullenia quinqueloba* Reuss – Cushman & Todd, p. 10, Pls. 2:5; 3:8. □1951 *Pullenia quinqueloba* Reuss – Marks, p. 69, Pl. 7:19a–b. □1957 *Pullenia quinqueloba* Reuss – Agip Mineraria, Pl. 44:10. □1958 *Pullenia quinqueloba* Reuss – Becker & Dusenbury, p. 27, Pl. 6:3a–b. □1959 *Pullenia quinqueloba* Reuss – Dieci, p. 87, Pl. 7:7a–b. □1971 *Pullenia quinqueloba* Reuss – Schnitker, p. 206, Pl. 10:11a–b. □1985 *Pullenia quinqueloba* Reuss – Kohl, p. 93, Pl. 32:6. □1985 *Pullenia quinqueloba* Reuss – Thomas, Pl. 4:2. □1990 *Pullenia quinqueloba* Reuss – Thomas *et al.*, Pl. 7:1; pl. 10:19. □1990 *Pullenia quinqueloba* Reuss – Ujiié, p. 43, Pl. 24:1a–b, 2, 3a–b, 4, 5.

*Description.* – Test free, planispiral, involute, compressed, slightly lobate, rounded to subrounded in side view. Chambers distinct, usually four to five in the last whorl, gradually increasing in size as added. Sutures distinct, slightly depressed, nearly radial. Wall calcareous, hyaline, smooth, finely perforate. Aperture a narrow interiomarginal slit extending to the umbilicus on either side, with a lip.

*Remarks.* – *Pullenia quinqueloba* is marked by a flattened test with lobate periphery, and usually 5–5.5 chambers.

*Distribution at Sites 502A and 503B.* – *Pullenia quinqueloba* is a common species at Site 502A, with relative abundances of 0–5%. It is rare with scattered occurrences at Site 503B.

## *Pullenia subcarinata* (d'Orbigny, 1839)

Fig. 24A–C

Synonymy. – □1839a *Nonionina subcarinata* n.sp. – d'Orbigny, p. 28, Pl. 5:23–24. □1932 *Pullenia subcarinata* (d'Orbigny) – Heron-Allen & Earland, pp. 403–404, Pl. 13:14–18. □1960 *Pullenia subcarinata* (d'Orbigny) – Barker, p. 174, Pl. 84:14–15.

*Description.* – Test free, planispiral, involute, compressed, periphery rounded or subrounded. Chambers distinct, 4–6 in the last whorl, increasing in size as added. Sutures distinct, slightly depressed. Wall calcareous, hyaline, smooth, finely perforate. Aperture a narrow interiomarginal slit that extending from one umbilicus to the other, with a lip.

*Remarks.* – *P. subcarinata* (d'Orbigny) differs from *P. quinqueloba* in possessing six chambers in the last whorl and a more inflated test.

*Ecology.* – *Pullenia subcarinata* has been reported from the upper middle bathyal zone of the Pacific Ocean (Ingle 1980) and the Gulf of Mexico (Pflum & Frerichs 1976). On the Ontong Java Plateau, *P. subcarinata* is found to be more abundant during postglacials (Burke *et al.* 1993).

*Distribution at Sites 502A and 503B.* – *Pullenia subcarinata* is rare with scattered occurrences at Site 503B. It occurs in most samples at Site 502A with relative abundances of 0–3%.

# Family Alabaminidae Hofker, 1951

# Genus *Gyroidina* d'Orbigny, 1826

## *Gyroidina altiformis* Stewart & Stewart, 1930

Fig. 24D–F

Synonymy. – □1930 *Gyroidina soldanii* d'Orbigny var. *altiformis* Stewart & Stewart, p. 67, Pl. 9:2a–c. □1949 *Gyroidina neosoldanii* Brotzen var. *acuta* – Boomgaart, p. 125, Pl. 14:1a–c. □1954 *Gyroidina soldanii* d'Orbigny var. *altiformis* – Parker, p. 527, Pl. 9:7–9. □1964 *Gyroidina altiformis* Stewart & Stewart – Leroy, p. 37, Pl. 7:7–9. □1966 *Gyroidina acuta* Boomgaart – Belford, p. 165, 167, Pl. 28:1–9, Text-fig. 21:6–7. □1976 *Gyroidina altiformis acuta* Boomgaart – Pflum & Frerichs, Pl. 4:8–9, Pl. 5:1. □1978 *Gyroidina altiformis* Stewart & Stewart – Wright, p. 714, Pl. 5:1–3. □1980 *Gyroidina altiformis* Stewart & Stewart– Ingle *et al.*, p. 138, Pl. 7:5–6. □1985 *Gyroidinoides altiformis* (Stewart & Stewart) – Kohl, pp. 95–96, Pl. 34:3a–c. □1985 *Gyroidinoides acutus* (Boomgaart) –

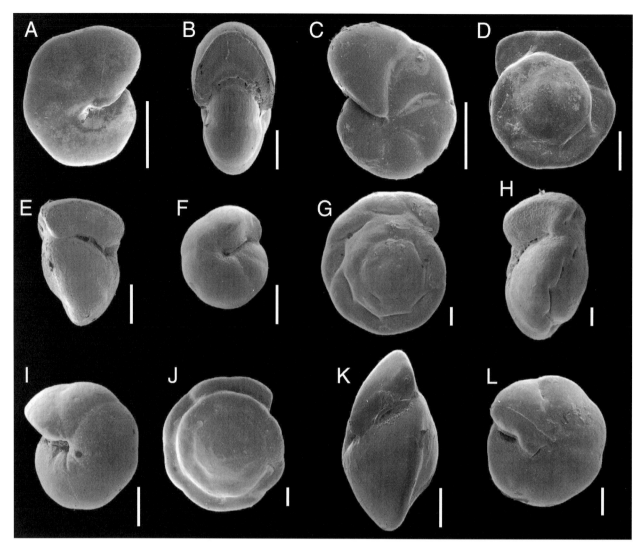

*Fig. 24.* Scale bar 100 µm. □A–C. *Pullenia subcarinata* (d'Orbigny); A, Side view, Hole 502A, 116.52 m; B, Edge view, Hole 502A, 116.52 m; C, Opposite side view, Hole 502A, 113.42 m. □D–F. *Gyroidina altiformis* Stewart & Stewart; D, Spiral view, Hole 502A, 41.62 m; E, Edge view, Hole 502A, 41.62 m; F, Umbilical view, Hole 502A, 106.53. □G–I. *Gyroidina neosoldanii* Brotzen; G, Spiral view, Hole 502A, 94.77 m; H, Edge view, Hole 502A, 94.77 m; I Umbilical view, Hole 502A, 103.63 m. □J–L. *Oridorsalis umbonatus* (Reuss); J, Spiral side, Hole 502A, 111.48 m; K, Edge view, Hole 502A, 111.48 m; L, Umbilical view, Hole 502A, 111.48 m.

Thomas, p. 676, Pl. 6:4–6. □1986 *Gyroidina acuta* Boomgaart – Kurihara & Kennett, Pl. 7:7. □1989 *Gyroidina altiformis* Stewart & Stewart – Hermelin, p. 81, Pl. 15:13–15.

*Description.* – Test free, trochospiral, planoconvex, spiral side flat to slightly convex, umbilical side convex with deep umbilicus. Circular in outline, peripheral edge subacute. Seven to ten chambers in the final whorl. Sutures distinct, limbate, slightly depressed and radial on the umbilical side, raised and oblique on the spiral side. Wall calcareous, hyaline, smooth, finely perforate. Aperture a low interiomarginal slit extending from periphery to umbilicus.

*Remarks.* – *Gyroidina altiformis* differs from *Gyroidina neosoldanii* in having raised, oblique sutures on the spiral side rather than flush, radial sutures.

*Ecology.* – Smith (1964) reported *G. altiformis* from the lower bathyal zone off the west coast of America, whereas Ingle *et al.* (1980) and Ingle (1980) observed it in the upper middle bathyal zone. *Gyroidina altiformis* is found to be more abundant during glacials than during interglacials on the Ontong Java Plateau (Burke *et al.* 1993).

*Distribution at Sites 502A and 503B.* – *Gyroidina altiformis* is a rare species with scattered occurrences at both Site 502A and Site 503B.

## *Gyroidina neosoldanii* Brotzen, 1936

Fig. 24G–I

*Synonymy.* – □1826 *Gyroidina soldanii* n.sp. – d'Orbigny, p. 278, modèles no. 36. □1936 *Gyroidina neosoldanii* sp.nov. – Brotzen, p. 158. □1953 *Gyroidina soldanii* d'Orbigny, 1826 – Phleger *et al.*, p. 41, Pl. 9:1–2. □1960 *Gyroidina neosoldanii* Brotzen, 1936 – Barker, p. 220, Pl. 107:6–7. □1964 *Gyroidina neosoldanii* Brotzen, 1936 – Leroy, p. 37, Pl. 7:4–6. □1965 *Gyroidina soldanii* d'Orbigny, 1826 – Todd, p. 19, Pl. 6:4. □1978 *Gyroidinoides soldanii* (Brotzen) – Lohmann, p. 29, Pl. 1:1–3. □1979 *Gyroidinoides soldanii* (Brotzen)– Corliss, p. 9, Pl. 5:4–6. □1980 *Gyroidina neosoldanii* Brotzen [*sic*] – Ingle *et al.*, p. 138, Pl. 7:10–11. □1981 *Gyroidina neosoldanii* Brotzen, 1936 – Resig, Pl. 8:5. □1985 *Gyroidina soldanii* d'Orbigny, 1826 – Hermelin & Scott, p. 220, Pl. 5:6–8. □1989 *Gyroidina neosoldanii* Brotzen, 1936 – Hermelin, p. 81, Pl. 15:16–18. □1990 *Gyroidina soldanii* d'Orbigny, 1826 – Thomas *et al.*, Pl. 10:22. □1990 *Gyroidinoides soldanii* (Brotzen) – Ujiié, p. 45, Pl. 25:1a–c, 2, 3a–c, 4a–c, 5a–c.

*Description.* – Test free, trochospiral, spiral side slightly convex, umbilical side strongly convex. Chambers have umbilical shoulders. Sutures distinct, radial or pointing slightly backwards, slightly depressed. Wall calcareous, perforate, granular in structure. Aperture a low interiomarginal slit in the middle of the apertural face and opening into the wide umbilicus.

*Distribution at Sites 502A and 503B.* – *Gyroidina neosoldanii* is a common species at both investigated sites with absolute abundances in the range of 0–18 individuals in each sample at Site 502A and 0–8 at Site 503B.

## Genus *Oridorsalis* Andersen, 1961

## *Oridorsalis umbonatus* (Reuss)

Fig. 24J–L

*Synonymy.* – □1884 *Truncatulina tenera* n.sp. – Brady, p. 665, Pl. 95:11. □1937 *Eponides umbonatus* (Reuss) – Chapman & Parr, p. 108. □1949 *Eponides umbonatus* (Reuss) – Bermúdez, p. 249, Pl. 17:22–24. □1951 *Gyroidina tenera* (Brady) – Hofker, p. 403, Text-figs. 279–280. □1952 *Eponides umbonatus* (Reuss) – Parker, p. 419, Pl. 6:13. □1953 *Eponides umbonatus* (Reuss) – Phleger *et al.*, p. 42, Pl. 9:9–10. □1954 *Eponides tenera* (Brady) – Cushman *et al.*, p. 359, Pl. 89:20. □1954 *Pseudoeponides tenera* (Brady) – Parker, p. 530, Pl. 9:20–21. □1958 *Pseudoeponides umbonatus* (Reuss) – Parker, p. 267, Pl. 3:30–32. □1960 *Eponides (?) tenera* (Brady) – Barker, p. 196, Pl. 95:11a–c. □1960 *Eponides umbonatus* (Reuss) – Barker, p. 216, Pl. 105:2a–c. □1964 *Pseudoeponides umbonatus* (Reuss) – Leroy, p. 39, Pl. 7:33–38. □1964 *Pseudoeponides umbonatus* (Reuss) – Smith, p. 43, Pl. 4:8a–c. □1965 *Oridorsalis umbonatus* (Reuss) – Todd., p. 23, Pl. 6:2. □1966 *Oridorsalis umbonatus* (Reuss) – Belford, pp. 172–173, Pl. 30:1–6. Text-figs. 22:4–5. □1966 *Pseudoeponides umbonatus* (Reuss) – Belford, p. 172, Pl. 30:1–6. □1966 *Oridorsalis umbonatus* (Reuss) – Todd, 1966, p. 29, Pl. 6:5; Pl. 13:5. □1971 *Oridorsalis tenera* (Brady) – Echols, p. 166, Pl. 15:3a–b. □1971 *Eponides tener tener* (Brady) – Herb, p. 298. □1973 *Oridorsalis umbonatus* (Reuss) – Douglas, Pl. 13:1–6, Pl. 24:9–12. □1976 *Oridorsalis tener umbonatus* (Reuss) – Pflum & Frerichs, p. 108, Pl. 6:5–7. □1978 □1976 *Oridorsalis tener tener* (Brady) – Pflum & Frerichs, Pl. 6:2–4. □1978 *Oridorsalis umbonatus* (Reuss) – Boltovskoy, Pl. 5:5–6. □1978 *Oridorsalis tener* (Brady) – Lohmann, p. 26, Pl. 4:5–7. □1978 *Oridorsalis umbonatus* (Reuss) – Lohmann, p. 26, Pl. 4:1–3. □1979 *Oridorsalis tener* (Brady) – Corliss, p. 9, Pl. 4:10–15. □1980 *Oridorsalis umbonatus* (Reuss) – Butt, Pl. 7:15, Pl. 7:21. □1980 *Oridorsalis tener* (Brady) – Ingle *et al.*, p. 142, Pl. 5:5–6. □1981 *Oridorsalis umbonatus* (Reuss) – Burke, Pl. 3:9–10. □1981 *Oridorsalis umbonatus* (Reuss) – Cole, p. 113, Pl. 14:8. □1981 *Oridorsalis umbonatus* (Reuss) – Resig, Pl. 8:8. □1984b *Oridorsalis umbonatus* (Reuss) – Boersma, Pl. 4:10–13. □1985 *Oridorsalis umbonatus* (Reuss) – Hermelin & Scott, p. 214, Pl. 5:10. □1985 *Oridorsalis umbonatus* (Reuss) – Kohl, p. 95, Pl. 33:6; Pl. 34:1–2. □1985 *Oridorsalis umbonatus* (Reuss) – McDougall, p. 396, Pl. 6:11. □1985 *Oridorsalis umbonatus* (Reuss) – Mead, p. 237, Pl. 5:8a–13. □1986 *Oridorsalis umbonatus* (Reuss) – Kurihara & Kennett, Pl. 6:11–13. □1989 *Oridorsalis umbonatus* (Reuss) – Hermelin, pp. 81–83, Pl. 16:1–5. □1990 *Oridorsalis umbonatus* (Reuss) – Schröder-Adams *et al.*, p. 34, Pl. 8:13. □1990 *Oridorsalis tenera* (Brady) – Thomas *et al.*, Pl. 10:20–21. □1990 *Oridorsalis umbonatus* (Reuss) – Thomas *et al.*, p. 227. □1990 *Oridorsalis umbonatus* (Reuss) – Ujiié, p. 48, Pl. 28:1a–c, 2a–c, 5a–c, 6; Text-fig. 4. □1991 *Oridorsalis umbonatus* (Reuss) – Scott & Vilks, p. 32, Pl. 4:4–5.

*Description.* – Test free, trochospiral, biconvex, peripheral margin carinate with a distinct keel, circular in side view. All chambers visible on spiral side, increasing gradually as added. Sutures radial and slightly curved on spiral side, strongly sinuosidal on umbilical side. Wall calcareous, hyaline, smooth, finely perforate. Aperture an interiomarginal slit extending from the umbilical area to the periphery, supplementary apertures on the spiralside at proximal end of the sutures of the last-formed chambers, similar apertures on the umbilical side are located along the suture between ultimate and penultimate chambers.

*Remarks.* – The morphological differences between *Oridorsalis umbonatus* and *Oridorsalis tener* are almost negligible. *Oridosalis umbonatus* has straight sutures on the spiral side, chambers of equal size in the last whorl, and a trochoidal cross-sectional outline, whereas *O. tener* has curved sutures, more rapid whorl expansion, and a more compressed outline (Lohmann, 1978; Corliss, 1979; Mead, 1985). I do not regard these forms as distinct species and therefore include *O. tener* in the synonymy of *O. umbonatus*. Hermelin (1989) contains a more extensive discussion of the *O. umbonatus – O. tener* group.

*Ecology. – Oridorsalis umbonatus* is an important cosmopolitan species of lower bathyal and abyssal faunas in the Indian Ocean (Corliss 1979), the Atlantic Ocean (Streeter 1973; Lohmann 1978; Streeter & Shackleton 1979), and the Antarctic (Uchio 1960) and is often associated with AABW (Corliss 1979). According to Mead (1985), this species occurs persistently in low abundances throughout the world's oceans, at 42–4,800 m depth. From surface sediments and piston cores in the Sulu Sea, Linsley *et al.* (1985) found that *O. umbonatus* is the most abundant species between 1,600 and 4,000 m water depth. Miao & Thunell (1993) found *O. umbonatus* abundant in the Sulu Sea at depths below 3,000 m and in sediment associated with relatively low organic-carbon content and deep pore water oxygen-penetration depths. *Oridorsalis umbonatus* is not found as an important species in the lower bathyal and abyssal faunas of either Pacific Ocean (Culp 1977) or the South China Sea (Miao & Thunell 1993), where the dominant species are agglutinated or dissolution-resistant hyaline forms. *Oridorsalis umbonatus* has been reported to prefer well-oxygenated, low-organic-carbon environments associated with regions of low surface productivity (Mackensen *et al.* 1985; Burke *et al.* 1993), whereas Woodruff (1985) and Woodruff & Savin (1989) found that *O. umbonatus* is more related to regions with high productivity in the surface area. However, observations by Rathburn & Corliss (1994) from the Sulu Sea indicate that *O. umbonatus* can be found in relatively low oxygen concentrations in sediments encompassing a wide range of organic-carbon values. Corliss & Chen (1988) listed this species as an epifaunal form, but a more recent study from the Sulu Sea by Rathburn & Corliss (1994) indicates that *O. umbonatus* should be categorized as a transitional infaunal species with distribution extending from 0 to 4–4.5 cm depth in the sediment.

*Distribution at Sites 502A and 503B. – Oridorsalis umbonatus* is one of the most common species at both sites. Absolute abundances are 0–47 individuals per sample at Site 502A and 0–26 at Site 503B.

# Family Osangulariidae Loeblich & Tappan, 1964

## Genus *Gyroidinoides* Brotzen, 1942

### *Gyroidinoides lamarckianus*, 1839
Fig. 25A–C

*Synonymy. –* □1839c *Rotalina lamarckiana* n.sp. – d'Orbigny, p. 131, Pl. 2:13–15. □1953 *Gyroidina lamarckiana* (d'Orbigny) – Phleger *et al.*, p. 41, Pl. 8:33–34. □1974 *Gyroidina lamarckiana* (d'Orbigny)– LeCalvez, p. 72, 74, Pl. 17:1–3. □1976 *Gyroidina orbicularis* d'Orbigny – Pflum & Frerichs, Pl. 5:5–7 (not *Gyroidina orbicularis* d'Orbigny, 1826). □1978 *Gyroidina lamarckiana* (d'Orbigny) – Boltovskoy, Pl. 4:14–15. □1979 *Gyroidinoides orbicularis* (d'Orbigny) – Corliss, p. 9, Pl. 5:1–3 (not *Gyroidina orbicularis* d'Orbigny, 1826).□1981 *Gyroidina lamarckiana* (d'Orbigny)– Burke, Pl. 3:7–8. □1981 *Gyroidina lamarckiana* (d'Orbigny) – Resig, Pl. 8:1–2. □1985 *Gyroidinoides* sp. A. Mead, p. 238, Pl. 5:1–3. □1985 *Gyroidinoides lamarckianus* (d'Orbigny) – Thomas, p. 677, Pl. 5:1–2. □1989 *Gyroidinoides lamarckianus* (d'Orbigny) – Hermelin, p. 83, Pl. 16:6, 10.

*Description. –* Test free, trochospiral, spiral side flat, umbilical side strongly convex, circular in outline, periphery rounded. Chambers indistinct, all visible on the spiral side. Sutures flush with surface, radial on spiral and umbilical side. Wall calcareous, hyaline, smooth, finely perforate. Aperture an interiomarginal slit in the apertural face midway between the periphery and umbilicus.

*Remarks. –* According to Hermelin (1989), the forms referred to as *Gyroidina orbicularis* by Phleger *et al.* (1953), Pflum & Frerichs (1976), and Corliss (1979) are probably representatives of this species.

*Ecology. –* According to Rathburn & Corliss (1994), the genus *Gyroidinoides* has intragenetic differences in the test morphology that include the variable distribution of pores over the umbilical side of the test, rounded peripheries, and variable spire height. *Gyroidinoides* is also related to infaunal or transitional (0–4 cm) microhabitat preference.

*Distribution at Sites 502A and 503B. – Gyroidinoides lamarckiana* is a common species at both sites. Absolute abundances are 0–9 individuals per sample at Site 502A and 0–7 at Site 503B.

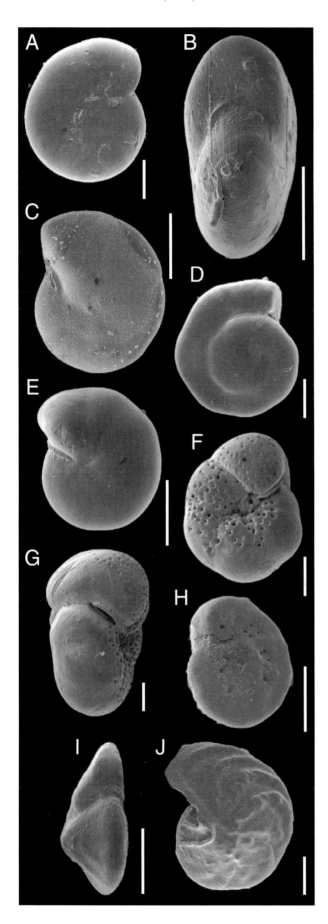

## *Gyroidinoides orbicularis* (d'Orbigny, 1826)

Fig. 25D–E

*Synonymy.* – □1826 *Gyroidina orbicularis* n.sp. – d'Orbigny, p. 278, no. 1, modèles no. 13. □1953 *Gyroidina orbicularis* d'Orbigny, 1826 – Phleger *et al.*, p. 41, Pl. 8:35–36. □1976 *Gyroidina orbicularis* d'Orbigny, 1826 – Pflum & Frerichs, p. 118, Pl. 5:5–7. □1979 *Gyroidina orbicularis* d'Orbigny, 1826 – Corliss, p. 9, Pl. 5:1–3. □1989 *Gyroidinoides orbicularis* (d'Orbigny) – Hermelin, pp. 83–84, Pl. 16:7–9.

*Description.* – Test free, trochoid, unequally biconvex; spiral side slightly convex, umbilical side strongly convex with slight depression, circular in outline. Chambers distinct, 8–12 in the last whorl, all visible on the spiral side. Sutures distinct, flush with surface, radial on spiral and umbilical sides. Wall calcareous, smooth, hyaline, finely perforate. Aperture an interiomarginal slit in the apertural face midway between the periphery and umbilicus, with a narrow lip.

*Ecology.* – In the Gulf of Mexico, Pflum & Frerichs (1976) reported that the bathymetric range of this species was from the upper depth limit of the upper bathyal zone down into abyssal depths, whereas it is found in the lower bathyal and abyssal zones of the Indian Ocean (Corliss 1979). *Gyroidinoides orbicularis* was also found in the lower bathyal zone off New Foundland (Cole 1981). It is reported by Schnitker (1980) from neritic depths on the North Carolina continental shelf.

*Distribution at Sites 502A and 503B.* – *Gyroidinoides orbicularis* is a relatively common species with abundances of 0–24 individuals per sample at Site 502A and 0–11 at Site 503B.

*Fig. 25.* Scale bar 100 µm. □A–C. *Gyroidinoides lamarckianus* (d'Orbigny); A, Spiral view, Hole 502A, 98.81 m; B, Edge view, Hole 502A, 103.63 m; C, Umbilical view, Hole 502A, 98.81 m. □D, E. *Gyroidinoides orbicularis* (d'Orbigny); D, Spiral view, Hole 502A, 114.45 m; □E Umbilical view, Hole 502A, 128.36 m. □F, G. *Anomalinoides globulosus* (Chapman & Parr); F, Umbilical view, Hole 502A, 64.51 m; G, Edge view, Hole 502A, 50.43 m. □H–J, *Cibicidoides bradyi* (Trauth); H, Umbilical view, Hole 502A, 88.84 m; I, Edge view, Hole 502A, 110.47 m; J, Spiral view, Hole 502A, 88.84 m.

# Family Anomalinidae Cushman, 1927

# Subfamily Anomalininae Cushman, 1927

# Genus *Anomalinoides* Brotzen, 1942

## *Anomalinoides globulosus* (Chapman & Parr, 1937)

Fig. 25F–G

*Synonymy.* – □1937 *Anomalina globulosa* sp.nov. – Chapman & Parr, p. 117, Pl. 9:27. □1953 *Anomalina* sp. Phleger *et al.*, p. 48, Pl. 10:26–28. □1960 *Anomalina globulosa* Chapman & Parr, 1937 – Barker, p. 194, Pl. 19:4–5. □1976 *Anomalinoides* sp. 1 Douglas – Resig, Pl. 4:10–11. □1978 *Anomalinoides* sp. – Lohmann, Pl. 2:13–15. □1981 *Anomalina globulosa* Chapman & Parr, 1937 – Resig, Pl. 8:11–12. □1985 *Anomalina globulosa* Chapman & Parr, 1937 – Hermelin & Scott, p. 203, Pl. 6:1–2. □1985 *Anomalinoides globulosus* (Chapman & Parr) – Thomas, Pl. 12:6–7. □1986 *Anomalinoides globulosa* (Chapman & Parr) – Boersma, Pl. 2:1–3. □1986 *Anomalinoides globulosus* (Chapman & Parr) – Kurihara & Kennett, Pl. 9:9. □1986 *Anomalinoides globulosus* (Chapman & Parr) – Van Morkhoven *et al.*, pp. 36–38, Pl. 9:1–3. □1989 *Anomalinoides globulosus* (Chapman & Parr) – Hermelin, pp. 84–85, Pl. 17:1; 5.

*Description.* – Test free, trochospiral, periphery rounded; spiral side strongly convex, umbilicus depressed. Chambers distinctly globular. Sutures depressed and slightly curved backwards. Wall coarsely perforated, giving test a rough appearance, especially on central parts. Aperture a low interiomarginal slit which begins at near the midpoint of the periphery and extends into the umbilicus, with a narrow lip.

*Ecology.* – According to Berggren *et al.* (1976) and Pflum & Frerichs (1976) *Anomalinoides globulosus* is a cosmopolitan species and is common from the middle bathyal to abyssal water depths.

*Distribution at Sites 502A and 503B.* – *Anomalinoides globulosus* is a rare species with scattered occurrences at both investigated sites.

# Genus *Cibicidoides* Thalmann, 1939

## *Cibicidoides bradyi* (Trauth, 1918)

Fig. 25H–J

*Synonymy.* – □1918 *Truncatulina bradyi* sp.nov. – Trauth, p. 235. □1942 *Cibicides bradyi* (Trauth) – Thalmann, p. 464. □1951 *Cibicides hyalina* Hofker, p. 359, Figs. 244–245. □1960 *Cibicides bradyi* (Trauth) – Barker, p. 196, Pl. 95:5a–c. □1964 *Eponides hyalinus* (Hofker) – Leroy, p. 37, Pl. 7:24–26. □1966 *Parelloides bradyi* (Trauth) – Belford, pp 100– 102, Pl. 11:10–19. □1971 *Cibicides bradyi* (Trauth) – Schnitker, p. 196, Pl. 10:12a–c. □1976 *Cibicides bradyi* (Trauth) – Pflum & Frerichs, Pl. 3:6–7. □1978 *Cibicides bradyi* (Trauth)– Boltovskoy, Pl. 3:6–8.

*Description.* – Test free, biconvex, periphery broadly rounded, 4–5 whorls visible on the spiral side, 9–11 chambers in the final whorl. Sutures slightly depressed. Wall calcareous, spiral side coarsely perforate, finely perforate on umbilical side. Aperture small and peripheral.

*Remarks.* – *Cibicidoides bradyi* is similar to *C. robertsonianus* but has a smaller size and lacks both an angular periphery and an imperforate keel. In addition, this species differs morphologically from many *Cibicidoides* species in that the periphery is rounded and the test may have a high spire with pores extending over most of the surface.

*Ecology.* – *Cibicidoides bradyi* is reported to be cosmopolitan deep-sea species (Uchio 1960; Streeter 1973; Lohmann 1978; Corliss 1979; Streeter & Shackleton 1979; Burke *et al.* 1993; Miao & Thunell 1996). In the southeast Indian Ocean, *C. bradyi* is an uncommon species found between 2,500 and 4,600 m (Corliss 1979). *Cibicidoides bradyi* has its upper depth limit in the upper middle bathyal zone in the western Pacific Ocean (Ingle & Keller 1980) and in the Gulf of Mexico (Pflum & Frerichs 1976). According to Corliss (1991), *C. bradyi* is an intermediate infaunal taxon (1–4 cm). A more recent study by Rathburn & Corliss (1994) from the Sulu Sea indicate that *C. bradyi* lives between 0 and 3 cm and is not restricted to the surface. In the Sulu Sea, *C. bradyi* increases in percentage with increasing water depth and decreasing organic-carbon values, which indicates that this species is able to make use of limited or alternate food resources (Rathburn & Corliss 1994). Miao & Thunell (1996) found that *C. bradyi* increased in abundance during glacial conditions in the Sulu Sea.

*Distribution at Sites 502A and 503B.* – *Cibicidoides bradyi* is a relative common species at Site 503B with absolute abundances of 0–9 individuals in each sample. Its distribution is more scattered at Site 503B.

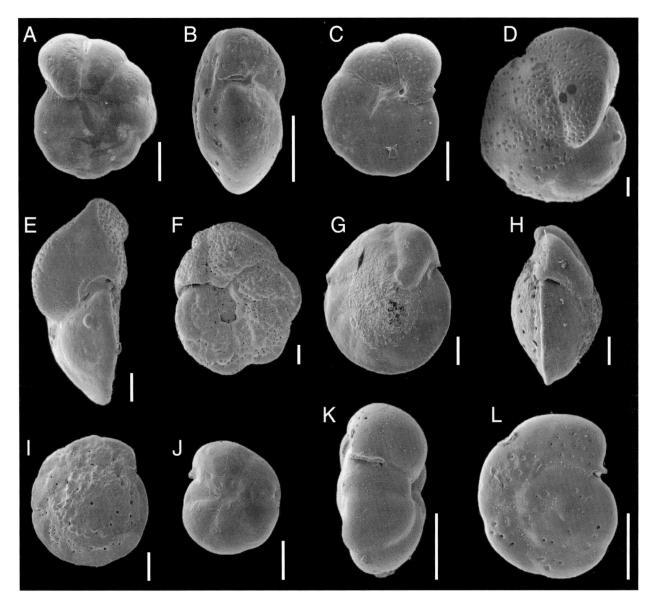

*Fig. 26.* Scale bar 100 μm. □A–C. *Cibicidoides kullenbergi* Parker; A, Spiral view, Hole 503B, 78.45 m; B, Edge view, Hole 503B, 104.54 m; C, Umbilical view, Hole 503B, 104.54 m. □D–F. *Cibicidoides lobatulus* (Walker & Jacob); D, Spiral view, Hole 502A, 91.85 m; E, Edge view, Hole 502A, 77.69 m; F, Umbilical view, Hole 502A, 77.69 m. □G–I. *Cibicidoides mundulus* (Bradyi, Parker & Jones); G, Spiral view, Hole 502A, 128.36 m; H, Edge view, Hole 502A, 128.36 m; I, Umbilical view, Hole 502A, 110.47 m. □J–L. *Cibicidoides robertsonianus* (Brady); J, Spiral view, Hole 502A, 110.47 m; K, Edge view, Hole 502A, 110.47 m; L, Umbilical view, Hole 502A, 110.47 m.

## *Cibicidoides kullenbergi* (Parker, 1953)

Fig. 26A–C

*Synonymy.* – □1953 *Cibicides kullenbergi* n.sp – Parker *in* Phleger *et al.*, p. 49, Pl. 11:7–8. □1978 *Cibicides kullenbergi* Parker *in* Phleger *et al.*, 1953 – Boltovskoy, Pl. 3:9–12. □1978 *Cibicidoides kullenbergi* (Parker) – Lohmann, p. 29, Pl. 2:5–7. □1978 *Cibicidoides kullenbergi* (Parker) – Wright, p. 713, Pl. 4:5–7. □1979 *Cibicidoides kullenbergi* (Parker) – Corliss, p. 10, Pl. 3:4–6. □1981 *Cibicidoides kullenbergi* (Parker) – Corliss & Honjo, p. 359, Pl. 2:1–16.

□1985 *Cibicidoides kullenbergi* (Parker) – Thomas, pp. 675–676, Pl. 8:1–2. □1978 *Cibicidoides* cf. *C. kullenbergi* (Parker) – Wright, p. 713, Pl. 4:8. □1985 *Cibicidoides* cf. *C. kullenbergi* (Parker) – Mead, p. 242, Pl. 6:6a–7b.

*Description.* – Test free, trochospiral, biconvex, composed of three whorls, periphery acute with a distinct, narrow keel. Chambers distinct, 11–14 in the last-formed whorl, increasing slowly in size, if at all, producing a circular outline with all chambers visible on spiral side. Sutures thin and flush with surface on both sides, curved. Wall calcareous, hyaline, smooth, finely perforate on the involute side, with large punctae rather widely spaced on each chamber on the evolute side. Aperture an interiomarginal equatorial arch, extending along the spiral sutures.

*Remarks.* – It has been debated whether *C. kullenbergi* (Parker) is a junior synonym of *Cibicidoides mundulus* (e.g., Van Morkhoven *et al.* 1986; Hermelin 1989). Several authors (e.g., Lohmann 1978; Wright 1978; Corliss 1979; Mead 1985) report an intermediate form between *C. kullenbergi* and *Cibicidoides wuellerstorfi*, which is found occasionally in deeper waters (about 4,000 m). The intermediate form differs from *C. kullenbergi* because it has only 7–9 chambers in the last whorl and a much less distinct or nonexistent keel. No umbilical plug exists in the intermediate form, which is characteristic for *C. kullenbergi*, while it is in all other respects clearly related to *C. kullenbergi*. Wright (1978) considered this form to be similar to *Cibicidoides robertsonianus* with regard to the shape and arrangement of the chambers but in other respects similar to *C. kullenbergi*. As a consequence, he named it *Cibicidoides* cf. *C. kullenbergi*. Lohmann (1978) regarded these intermediate forms as *C. kullenbergi*, while Corliss (1979) assumed that they represent ecophenotypic variation and referred them to *C. wuellerstorfi*. Thomas (1985) concluded that *C. kullenbergi* is closely related to *C. mundulus*, but referred her specimens to *C. kullenbergi* if they had a keel and if the sutures on the involute side are curved strongly and thus join the keel tangentially. Generally, *C. kullenbergi* has more chambers per whorl (about 13) than the other *Cibicidoides* species (9–11) (Thomas 1985). Hermelin (1989) included *Cibicidoides kullenbergi* in *C. mundulus*. On the basis of the present investigation, I have found *C. kullenbergi* to be sufficiently different from *C. wuellerstorfi* and *C. mundulus* to warrant its recognition as a separate species.

*Ecology.* – *Cibicidoides kullenbergi* is regarded as an indicator species of NADW in the Atlantic (Lohmann 1978). According to Pflum & Frerichs (1976), *C. kullenbergi* has its upper depth limits within the middle bathyal in the Gulf of Mexico. According to Corliss (1985, 1991) and Corliss & Chen (1988), *C. kullenbergi* is characterized as epifaunal and has its abundance maxima in the 0–1 cm interval. In the equatorial Pacific Ocean, Burke *et al.* (1993) reported that *C. kullenbergi* may prefer environment with low productivity.

*Distribution at Sites 502A and 503B.* – *Cibicidoides kullenbergi* is a common species with absolute abundances of 0–27 individuals per sample at Site 502A and 0–13 at Site 503B.

## *Cibicidoides lobatulus* (Walker & Jacob, 1798)

Fig. 26D–F

*Synonymy.* – ☐1798 *Nautilus lobatulus* sp.nov. – Walker & Jacob, p. 642, Pl. 14:36. ☐1884 *Truncatulina lobatula* (Walker & Jacob) – Brady, p. 660, Pl. 92:10, Pl. 93:1, 4–5, Pl. 115:4–5. ☐1953 *Cibicides lobatulus* (Walker & Parker) – Phleger *et al.*, p. 49, Pl. 11:9, 14. ☐1960 *Cibicides lobatulus* (Walker & Parker) – Barker, p. 190, Pl. 92:10, p. 192, Pl. 93:1; 4–5, p. 238, Pl. 115:4–5. ☐1969 *Cibicides lobatulus* (Walker & Parker) – Vilks, p. 50, Pl. 3:17a–b. ☐1978 *Cibicides lobatulus* (Walker & Parker) – Wright, p. 713. ☐1979 *Cibicides lobatulus* (Walker & Parker) – Corliss, p. 10, Pl. 3:7–9. ☐1990 *Cibicides lobatulus* (Walker & Parker) – Schröder-Adams *et al.*, p. 24. ☐1990 *Cibicides lobatulus* (Walker & Parker) – Thomas *et al.*, Pl. 10:4. ☐1991 *Cibicides lobatulus* (Walker & Parker) – Scott & Vilks, p. 28.

*Description.* – Test free, trochospiral, biconvex. Chambers distinct, lobe-shaped, irregularly shaped in some specimens, 8–10 in the final whorl, increasing rapidly in size, all chambers visible on spiral side. Sutures distinct, slightly curved, flush with surface. Wall calcareous, hyaline, smooth, coarsely perforate on spiral side, umbilical side imperforate. Aperture an interiomaginal arch with lip, continuing along the spiral suture.

*Remarks.* – *Cibicidoides lobatulus* generally has a small test with chambers that rapidly increase in size. It is assigned to *Cibicidoides* because of its weakly biconvex test and the presence of pores on the spiral side. In some specimens, the lobe-shaped chambers are irregularly shaped.

*Ecology.* – *Cibicidoides lobatulus* is found in the intertidal zones in the Gulf of Mexico as well as in many other areas (Pflum & Frerichs 1976). In the southeast Indian Ocean, *C. lobatulus* is found in low frequences at 2,500–3,800 m (Corliss 1979).

*Distribution at Sites 502A and 503B.* – *Cibicidoides lobatulus* is rare with scattered occurrences at Site 502A. It is absent at Site 503B.

## *Cibicidoides mundulus* (Brady, Parker & Jones, 1888)

Fig. 26G–I

*Synonymy.* – ☐1888 *Truncatulina mundula* n.sp. – Brady, Parker & Jones, p. 228, Pl. 45:25. ☐1960 *Cibicidoides mundulus* (Brady, Parker & Jones) – Barker, p. 196, Pl. 95:6a–c. ☐1981 *Cibicidoides mundulus* (Brady, Parker & Jones) – Resig, Pl. 8:15, 18. ☐1985 *Cibicidoides mundulus* (Brady, Parker & Jones) – Thomas, Pl. 8:5–6. ☐1986 *Cibicidoides mundulus* (Brady, Parker & Jones) –Kurihara & Kennett,

Pl. 8:7–9. □1986 *Cibicidoides mundulus* (Brady, Parker & Jones) – Van Morkhoven *et al.*, pp. 65–67, Pl. 21:1a–c. □1989 *Cibicidoides mundulus* (Brady, Parker & Jones) – Hermelin, p. 86, Pl. 17:9–11. □1990 *Cibicidoides mundulus* (Brady, Parker & Jones) – Ujiié, pp. 51–52, Pl. 30:1a–c, 2a–c, 3a–c.

*Description.* – Test free, trochospiral, biconvex, periphery angular with a narrow, thickened keel. Chambers distinct, increasing gradually in size as added, all chambers visible on the spiral side, 9–12 in the last whorl. Sutures slightly curved on the umbilical side, oblique on the spiral side. Wall calcareous, hyaline, smooth, coarsely perforate. Aperture an interiomarginal equatorial arch, extending along the spiral suture the length of one or two chambers.

*Ecology.* – Rathburn & Corliss (1994) reported that *C. mundulus* shows differences between the adult and juvenile assemblages in the Sulu Sea, where adults live primarily in the 0–1 cm depth interval and juveniles in deeper (1.5–6 cm) sediments. According to Rathburn & Corliss (1994), this pattern may reflect microhabitat differences between adults and juveniles for this species.

*Distribution at Sites 502A and 503B.* – *Cibicidoides mundulus* is a relatively common species at Site 502A, where absolute abundances fluctuate between 0 and 24 individuals per sample. It is rare with scattered occurrences at Site 503B.

## *Cibicidoides robertsonianus* (Brady, 1884)

Fig. 26J–L

*Synonymy.* – □1884 *Truncatulina robertsoniana* n.sp. – Brady, p. 664, Pl. 95:4. □1931 *Cibicides robertsoniana* (Brady) – Cushman, p. 121, Pl. 23:6a–c. □1951 *Cibicides robertsoniana* (Brady) – Phleger & Parker, p. 31, Pl. 16:10a–b, 11a–b, 12a–b, 13a–b. □1953 *Cibicides robertsonianus* (Brady) – Phleger *et al.*, p. 50, Pl. 11:15–17. □1954 *Cibicides robertsoniana* (Brady) – Parker, p. 543, Pl. 13:2, 5. □1957 *Cibicides robertsoniana* (Brady) – Agip Mineraria, Pl. 52:3a–c. □1976 *Cibicidoides robertsonianus* (Brady) – Pflum & Frerichs, Pl. 3:3–5. □1976 *Cibicides robertsoniana* (Brady) – Todd & Low, p. 22, Pl. 8:2a–c. □1985 *Cibicidoides robertsonianus* (Brady) – Thomas, Pl. 10:5–6. □1986 *Cibicidoides robertsonianus* (Brady) – Van Morkhoven *et al.*, pp. 41–43, Pl. 11:1a–c. □1989 *Cibicidoides robertsonianus* (Brady) – Hermelin, pp. 86–87.

*Description.* – Test free, trochospiral, plano-convex; peripheral margin rounded. Chambers increasing gradually in size as added; all chambers visible on the spiral side, almost involute on the umbilical side. Sutures distinct, slightly curved, flush with surface. Wall calcareous, hyaline, smooth; spiral side coarsely perforate, umbilical side

imperforate. Aperture an interiomarginal arch with a distinct lip, extending along the spiral suture.

*Remarks.* – *Cibicidoides robertsonianus* has a larger test, less rounded periphery, and imperforate umbilical side compared to *Cibicidoides bradyi*.

*Ecology.* – In the Gulf of Mexico, *Cibicidoides robertsonianus* is found to have the same upper depth limit as *Cibicidoides bradyi*, which is near the upper boundary of the upper middle bathyal zone (Pflum & Frerichs 1976). *Cibicidoides robertsonianus* is characterized as epifaunal and has its abundance maxima in the 0–1 cm interval (Corliss 1985, 1991; Corliss & Chen 1988).

*Distribution at Sites 502A and 503B.* – *Cibicidoides robertsonianus* is a relatively common species at Site 502A with relative abundances 0–27 individuals per sample. It exhibits more scattered occurrences at Site 503B.

## *Cibicidoides wuellerstorfi* (Schwager, 1866)

Fig. 27A–C

*Synonymy.* – □1866 *Anomalina wüllerstorfi* n.sp. – Schwager, p. 258, Pl. 7:105–107. □1884 *Truncatulina wuellerstorfi* (Schwager) – Brady, p. 662, Pl. 93:8–9. □1915 *Truncatulina wuellerstorfi* (Schwager) – Cushman, p. 34, Pl. 12:3a–c. □1921 *Truncatulina wuellerstorfi* (Schwager) – Cushman, pp. 314–315, Pl. 64:1a–c. □1929b *Planulina wuellerstorfi* (Schwager) – Cushman, p. 104, Pl. 15:1–2. □1953 *Planulina wuellerstorfi* (Schwager) Phleger *et al.*, p. 49, Pl. 11:1–2. □1958 *Cibicides wuellerstorfi* (Schwager) – Parker, p. 275, Pl. 4:41–42. □1960 *Planulina wuellerstorfi* (Schwager) – Barker, p. 192, Pl. 93:9a–c. □1964 *Cibicides wuellerstorfi* (Schwager) – Akers & Dorman, pp. 31–32, Pl. 15:16–17. □1964 *Cibicides wuellerstorfi* (Schwager) – Leroy, p. 45, Pl. 8:25–26. □1964 *Cibicidoides wuellerstorfi* (Schwager) – Parker, pp. 624–625, Pl. 100:29. □1965 *Planulina wuellerstorfi* (Schwager) – Todd, p. 51, Pl. 23:3–5. □1966 *Planulina wuellerstorfi* (Schwager) – Belford, pp. 120–121, Pl. 20:1–6. □1973 *Cibicides wuellerstorfi* (Schwager) – Douglas, Pl. 18:7–9, Pl. 25:15–16. □1976 *Cibicides wuellerstorfi* (Schwager) – Pflum & Frerichs, Pl. 4:2–4. □1978 *Cibicides wuellerstorfi* (Schwager) – Boltovskoy, Pl. 3:19–21. □1978 *Planulina wuellerstorfi* (Schwager) – Lohmann, p. 26, Pl. 2:1–3. □1979 *Planulina wuellerstorfi* (Schwager) – Corliss, pp. 7–8, Pl. 2:13–16. □1981 *Planulina wuellerstorfi* (Schwager) – Cole, p. 103, Pl. 12:9. □1981 *Cibicidoides wuellerstorfi* (Schwager) – Resig, Pl. 8:16–17. □1984 *Planulina wuellerstorfi* (Schwager) – Murray, Pl. 2:16–18. □1985 *Planulina wuellerstorfi* (Schwager) – Hermelin & Scott, p. 216, Pl. 4:12–13. □1985 *Cibicidoides wuellerstorfi* (Schwager) – Mead, p. 240, Pl. 6:1a–2. □1985 *Cibicidoides wuellerstorfi*

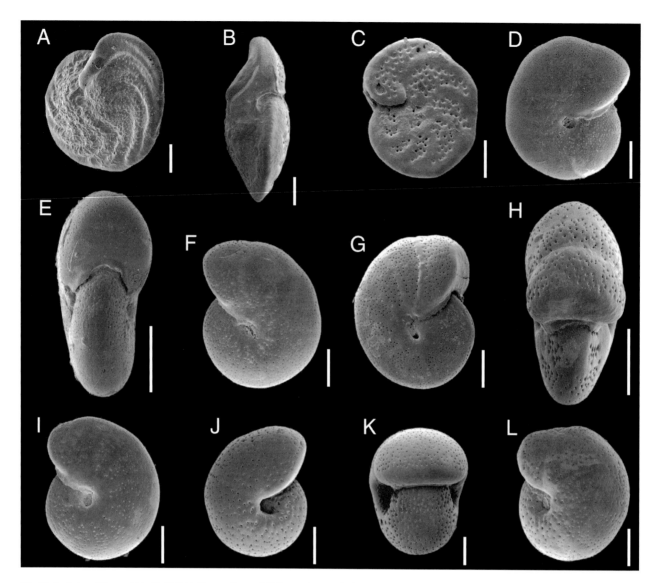

*Fig. 27.* Scale bar 100 µm. □A–C. *Cibicidoides wuellerstorfi* (Schwager); A, Spiral view, Hole 502A, 104.54 m; B, Edge view, Hole 502A, 104.54 m; C, Umbilical view, Hole 502A, 104.54 m. □D–F. *Melonis affinis* (Reuss); D, Side view, Hole 502A, 97.80 m; E, Edge view, Hole 502A, 97.80 m; F, Opposite side view, Hole 503B, 58.00 m. □G–I. *Melonis barleeanum* (Williamson); G, Side view, Hole 502A, 105.58 m; H, Edge view, Hole 502A, 105.58 m; I, Opposite side view, Hole 502A, 105.58 m. □J–L. *Melonis pompilioides* (Fichtel & Moll); J, Side view, Hole 502A, 65.51 m; K, Edge view, Hole 502A, 65.51 m; L, Opposite side view, Hole 502A, 65.51 m.

(Schwager) – Thomas, Pl. 11:1–4. □1986 *Cibicidoides wuellerstorfi* (Schwager) – Kurihara & Kennett, Pl. 9:4–6. □1986 *Planulina wuellerstorfi* (Schwager) – Van Morkhoven *et al.*, pp. 48–50, Pl. 14:1–2. □1989 *Cibicidoides wuellerstorfi* (Schwager) – Hermelin, p. 87. □1990 *Cibicidoides wuellerstorfi* (Schwager) – Thomas *et al.*, Pl. 4:9; Pl. 7:5.

*Description.* – Test free, plano-convex; periphery with a distinct keel. Sutures strongly recurved on both sides. Wall calcareous, coarsely and densely perforate. Aperture a low arch at the base of the final chamber.

*Remarks.* – *Cibicidoides wuellerstorfi* has commonly been placed in to the genus *Planulina*. I share the standpoint of

Loeblich & Tappan (1964), who concluded that this foraminifer belongs to the genus *Cibicidoides*. This species has a plano-convex test, where the sutures are strongly curved on the evolute side.

*Ecology.* – *Cibicidoides wuellerstorfi* is an important deep-sea species; it is cosmopolitan and most abundant at bathyal to abyssal depth. It has previously been interpreted as an indicator of NADW (Uchio, 1960; Lohmann, 1978; Schnitker, 1980; Hermelin, 1989, among others). According to Mackensen *et al.* (1985), *C. wuellerstorfi* lives predominantly at 2,000–3,000 m water depth, is adapted to deep-sea environments with relatively high food supply, and tolerates relatively low oxygen contents of the interstitial water. *Cibicidoides wuellerstorfi* has been found in the upper 1 cm in the Atlantic (Corliss 1985, 1991) as well in the Sulu Sea (Rathburn & Corliss 1994). According to Lutze & Thiel (1989), *C. wuellerstorfi* prefers a microhabitat elevated above the sediment–water interface. This microhabitat preference manifests itself in isotopic signatures of *C. wuellerstorfi* that are closely correlated with those of overlying bottom-waters (Belanger *et al.* 1981; Graham *et al.* 1981; Zahn *et al.* 1986; and Grossman, 1987). Miao & Thunell (1993) found *C. wuellerstorfi* to be abundant at 1,500–3,200 m water depths in the South China Sea and at 1,000–2,200 m in the Sulu Sea. In the western Pacific, *C. wuellerstorfi* is abundant below 3,500 m and associated with Pacific Deep Water (PDW) (Culp 1977; Walch 1978). There are conflicting reports whether the abundance of *C. wuellerstorfi* related to regions with high or low productivity; it has been found to become relatively more abundant in sediments with high organic-carbon content (Boersma 1985; Mackensen *et al.* 1985; Woodruff & Savin 1989; Loubere 1991) but has also been reported to respond positively to low organic-carbon environments by (Altenbach 1985; Caralp 1988; Lutze & Thiel 1989; and Burke *et al.* 1993). *Cibicidoides wuellerstorfi* is abundant at 1,500–3,200 m water depths in the South China Sea, and at 1,000–2,200 m in the Sulu Sea (Miao & Thunell 1993). In the western Pacific, *C. wuellerstorfi* is abundant below 3,500 m and associated with Pacific Deep Water (Culp 1977; Walch 1978).

*Distribution at Sites 502A and 503B.* – *Cibicidoides wuellerstorfi* is a common species at both sites, with absolute abundances of 0–155 individuals per sample at Site 502A and 0–21 at Site 503B.

## Genus *Melonis* de Montfort, 1808

### *Melonis affinis* (Reuss, 1851)

Fig. 27D–F

*Synonymy.* – □1851a *Nonionina affinis* n.sp. – Reuss, p. 72, Pl. 5:32a–b. □1858 *Nonionina barleeana* sp.nov. – Williamson, p. 32, Pl. 3:68–69. □1880 *Nonionina formosa* n.sp. – Seguenza, p. 63, Pl. 7:6. □1930 *Nonion planatum* sp.nov. – Cushman & Thomas, p. 37, Pl. 3:5. □1936b *Nonion nicobarensis* n.sp. – Cushman, p. 67, Pl. 12:9a–b. □1948 *Nonion affine* (Reuss) – Renz, p. 148, Pl. 6:3a–b. □1960 *Nonion parkerae* n.sp. – Uchio, p. 60, Pl. 4:9–10. □1964 *Nonion affine* (Reuss) – Smith, p. 41, Pl. 4:1a–b. □1966 *Melonis affinis* (Reuss) – Belford, p. 184, Pl. 31:1–4. □1978 *Nonion affine* (Reuss) – Boltovskoy, p. 162, Pl. Pl. 5:1–2. □1985 *Melonis affinis* (Reuss) – Kohl, p. 100, Pl. 36:4a–d.

*Description.* – Test free, planispiral, biconvex, compressed; periphery rounded, nearly circular in side view.

Chambers distinct, 10–11 in the last whorl, gradually increasing in size as added. Sutures limbate, flush with surface, gently curved, fused into an imperforate ring around the umbilicus. Wall calcareous, hyaline, smooth, coarsely perforate, except for the imperforate umbilical rim, umbilical flaps, and apertural face. Aperture a low-arched interiomarginal equatorial slit, extending laterally from one umbilicus to the other.

*Remarks.* – The topotypes of *Melonis affinis* from the Eocene of Germany were examined by Boltovskoy (1958). He noted that there were 10–11 chambers in the last whorl and that the sutures were limbate and became wider near the umbilicus, where they formed a 'circle of varying size at the circumference'. These observations broadened the description of Reuss (1851) and support the inclusion of many additional species in the synonymy of *M. affinis*.

*Ecology.* – *Melonis affinis* is a transitional infaunal taxon with a distribution from 0 to 4 cm (Rathburn & Corliss 1994). In the South China Sea, *M. affinis* has been found to be abundant at water depths shallower than 1,500 m, and it is also common between 2,500 m and 3,500 m (Miao & Thunell 1993).

*Distribution at Sites 502A and 503B.* – *Melonis affinis* is a relatively common species at Site 502A, with absolute abundances between 0 and 31 individuals in each sample. It is rare with scattered occurrences at Site 503B.

### *Melonis barleeanum* (Williamson, 1858)

Fig. 27G–I

*Synonymy.* – □1858 *Nonionina barleeana* n.sp. – Williamson, p. 32, Pl. 3:68–69. □1884 *Nonionina umbilicatula* (Montagu) – Brady, p. 726, Pl. 109:8–9. □1914 *Nonionina umbilicatula* (Montagu) – Cushman, pp. 24–25, Pl. 17:1. □1930 *Nonion barleeanum* (Williamson) – Cushman, p. 11, Pl. 4:5. □1939 *Nonion barleeanum* (Williamson) – Cushman, p. 23, Pl. 6:11. □1952 *Nonion barleeanum* (Williamson) – Phleger, p. 85, Pl. 14:6. □1952 *Anomalinoides barleeanum* (Williamson) var. *zaandamae* Voorthuysen, p. 681. □1953 *Nonion zaandamae* (Voorthuysen) – Loeblich & Tappan, p. 87, Pl. 16:11–12. □1953 *Nonion barleeanum* (Williamson) – Phleger *et al.*, p. 30, Pl. 6:4. □1960 *Gavelinonion barleeanum* (Williamson) – Barker, p. 224, Pl. 109:8–9. □1964 *Melonis zaandamai* [*sic*] (Williamson) – Loeblich & Tappan, p. C761–C763. □1976 *Melonis barleeanus* (Williamson) – Pflum & Frerichs, Pl. 7:5–6. □1978 *Melonis barleeanum* (Williamson) – Wright, p. 715, Pl. 6:4. □1979 *Melonis barleeanum* (Williamson) – Corliss, p. 10, 12, Pl. 5:7–8. □1980 *Melonis barleeanum* (Williamson) – Thompson, Pl. 7:4a–b. □1984b *Nonion barleanum* [*sic*] (Williamson) – Boersma, Pl. 3:11–13. □1984 *Melonis barleeanus* (Williamson) –

Murray, Pl. 2:8–9. □1985 *Nonion barleeanum* (Williamson) – Hermelin & Scott, p. 212, 214, Pl. 5:2. □1985 *Melonis barleeanus* (Williamson) – Thomas, Pl. 12:3. □1986 *Melonis barleeanum* (Williamson) – Kurihara & Kennett, Pl. 9:10–11. □1989 *Melonis barleeanum* (Williamson) – Hermelin, p. 88, Pl. 17:12. □1990 *Melonis barleeanum* (Williamson) – Thomas *et al.*, Pl. 10:12. □1990 *Melonis barleeanum* (Williamson) – Ujiié, p. 52, Pl. 29:4a–c. □1991 *Nonion barleeanum* (Williamson) – Scott & Vilks, p. 30, Pl. 4:6–7.

*Description.* – Test free, planispiral, involute, compressed, deeply biumbilicate, periphery rounded. Chambers distinct, usually more than ten in the final whorl. Sutures gently curved, flush with surface. Wall calcareous, hyaline, smooth, coarsely perforate, except for the apertural face. Aperture an interiomarginal slit that extends to umbilicus on both sides.

*Remarks.* – Several authors have placed this species in the genus *Nonion*, but due to the fact that the umbilical region is open rather than closed, it has been retained here in genus *Melonis*. *Melonis barleeanum* is similar to *M. pompilioides*, but is more compressed and more finely perforate.

*Ecology.* – Pflum & Frerichs (1976) found this species in the neritic zone and into the middle bathyal zone in the Gulf of Mexico. *Melonis barleeanum* prefers a fine-grained substrate with a rather high organic-carbon content (Qvale & van Weering 1985). It is generally considered as a bathyal and infaunal species (Corliss and Chen 1988).

*Distribution at Sites 502A and 503B.* – *Melonis barleeanum* is a relatively common species at Site 502A with absolute abundances between 0 and 23 individuals in each sample. It has more scattered occurrences at Site 503B.

## *Melonis pompilioides* (Fichtel & Moll, 1798)

Fig. 27J–L

*Synonymy.* – □1798 *Nautilus pompilioides* n.sp. – Fichtel & Moll, p. 31, Pl. 2:A–C. □1826 *Nonionina umbilicatulata* n.sp. – d'Orbigny, p. 293, Pl. 15:10–12. □1826 *Nonionina pompilioides* (Fichtel & Moll) – d'Orbigny, p. 294, no. 15. □1884 *Nonionina pompilioides* (Fichtel & Moll) – Brady, p. 727, Pl. 109:10–11. □1914 *Nonionina pompilioides* (Fichtel & Moll) – Cushman, pp. 25–26, Pl. 17:2. □1930 *Nonion pompilioides* (Fichtel & Moll) – Cushman, p. 4, Pl. 1:7–11, Pl. 2:1–2. □1946 *Nonion pompilioides* (Fichtel & Moll) – Cushman, p. 6, Pl. 1:1–2. □1951 *Nonion pompilioides* (Fichtel & Moll) – Phleger & Parker, p. 11, Pl. 5:19–20. □1953 *Nonion pompilioides* (Fichtel & Moll) – Phleger *et al.*, p. 30, Pl. 6:7–8. □1958 *Melonis pompilioides* (Fichtel

& Moll) – Voloshinova, p. 158, Pl. 3:8–9. □1958 *Melonis sphaeroides* n.sp. –Voloshinova, p. 153, Pl. 3:1a–b. □1960 *Nonion* (?) *pompilioides* (Fichtel & Moll) – Barker, p. 224, Pl. 109:10–11. □1966 *Melonis pompilioides* (Fichtel & Moll) – Belford, pp. 183–184, Pl. 30:17–20. □1973 *Melonis pompilioides* (Fichtel & Moll) – Douglas, Pl. 9:8–9. □1976 *Melonis pompilioides* (Fichtel & Moll) – Pflum & Frerichs, Pl. 7:7–8. □1978 *Nonion pompilioides* (Fichtel & Moll) – Boltovskoy, Pl. 5:3a–b. □1978 *Melonis pompilioides* (Fichtel & Moll) – Lohmann, p. 29, Pl. 1:12–13. □1979 *Melonis pompilioides* (Fichtel & Moll) – Corliss, p. 12, Pl. 5:9–10. □1980 *Melonis pompilioides* (Fichtel & Moll) – Ingle *et al.*, Pl. 9:14–15. □1980 *Melonis pompilioides* (Fichtel & Moll) – Keller, 1980, Pl. 3:11–12. □1980 *Melonis pompilioides* (Fichtel & Moll) – Thompson, Pl. 7:4a–b. □1984 *Melonis pompilioides* (Fichtel & Moll)– Murray, Pl. 2:10–11. □1985 *Melonis pompilioides* (Fichtel & Moll)– Hermelin & Scott, p. 212, Pl. 6:5. □1985 *Melonis sphaeroides* Voloshinova, 1958 – Mead, p. 242, 244, Pl. 7:3a–b. □1985 *Melonis pompilioides* (Fichtel & Moll) – Thomas, 1985, Pl. 12:1–2. □1986 *Melonis pompilioides* (Fichtel & Moll) – Kurihara & Kennett, Pl. 9:7–8. □1986 *Melonis pompilioides* (Fichtel & Moll) – Van Morkhoven *et al.*, pp. 72–80, Pl. 23A:1–2, Pl. 23C:1a–d; Pl. 23D:1a–d; Pl. 23e;1a–c. □1989 *Melonis pompilioides* (Fichtel & Moll) – Hermelin, pp. 88–89, Pl. 17:13–14. □1990 *Melonis pompilioides* (Fichtel & Moll) – Thomas *et al.*, Pl. 10:16.

*Description.* – Test free, planispiral, involute, periphery broadly rounded; eight to ten chambers in the final whorl. Wall calcareous, hyaline, smooth, coarsely perforate. Sutures slightly curved, flush with surface. Aperture a low arch that extends to umbilicus on both sides.

*Remarks.* – *Melonis pompilioides* is a cosmopolitan, deeper-water species, which has been widely recorded in the literature. It has been referred to the genera *Nonion* and *Melonis* under many species names: *M. soldanii* (d'Orbigny), *M. sphaeroides* (Voloshinova), *Nonionina umbilicatula* d'Orbigny, and *Nonion halkyardi* Cushman. I have followed Hermelin's (1989) proposal that these various forms represent ecophenotypic variants that should be regarded as junior synonyms of *M. pompilioides*.

*Ecology.* – *Melonis pompilioides* is found in the bathyal zone of the Gulf of Mexico (Pflum & Frerichs 1976). Ingle (1980) and Ingle & Keller (1980) observed this species in the lower bathyal zone in the eastern Pacific. On the Ontong Java Plateau, this species is abundant during Pliocene, with a relative abundance up to 7.1% (Hermelin 1989), and it is also abundant during glacials (Burke *et al.* 1993).

*Distribution at Sites 502A and 503B.* – *Melonis pompilioides* is an abundant species, especially above 83.03 m, at Site 502A. It is abundant down to approximately 72.03 m at Site 503B.

# References

Agip Mineraria 1957: Foraminiferi Padani (*Terziario e Quaternario*). 52 pls. Milan.

Akers, W.H. & Dorman, J.H. 1964: Pleistocene foraminifera of the Gulf Coast. *Tulane Studies in Geology 3*, 9–93.

Alliata, E.D.N. 1946: Contributo alla Conoscenza della Stratigrafia del Pliocene e del Calabriano nella Regione di Rovigo. *Rivesta Italiana di Paleontologia 52*, 19–36.

Alliata, E.D.N. 1947: Sull'esistenza del Calabriano e del Siciliano, Rivelata dai Microfossili, nel Suttosuolo della Pianura Lodigiana (Milano). *Rivesta Italiana di Paleontologia 53*, 19–24.

Allmon, W.D., Rosenberg, G., Portell, R.W. & Schindler, K.S. 1993: Diversity of Atlantic Coastal Plain mollusks since the Pliocene. *Science 260*, 1626–1629.

Andersen, H.V. 1961: Genesis and Paleontology of the Mississippi River mudlumps, lower Mississippi River Delta, Louisiana. *Louisiana Geological Society, Bulletin 35*, 1–208.

[Altenbach, A.V. 1985: Die Biomasse der benthischen Foraminiferen. Auswertungen von 'Meteor'-Expeditionen im Östlichen Nordatlantic. 167 pp. Ph.D. diss., University of Kiel.]

Andersson, J.B. 1975: Ecology and distribution of foraminifera in the Weddell Sea of Antarctica. *Micropaleontology 21*, 69–96.

Asano, K. 1950: Part 3, Textulariidae. *In* L.W. (ed.): *Illustrated Catalogue of Japanese Tertiary Smaller Foraminifera*, 1–7.

Asano, K. 1951: Part 14, Rotaliidae. *In* L.W. (ed.): *Illustrated Catalogue of Japanese Tertiary Smaller Foraminifera 6*, 1–20.

Asano, K. 1956: The foraminifera from the adjacent sea of Japan, collected by the S.S. Soyo-Maru, 1922–1930. Part 1 Nodosariidae. Tohoku University, *Science Reports, 2nd Series (Geology), 27*, 1–55.

Asano, K. 1958: The foraminifera from the adjacent sea of Japan, collected by the S.S. Soyo-Maru, 1922–1930. Part 4 Buliminidae. Tohoku University, *Science Reports, 2 nd Series (Geology), 29*, 1–41.

Backman, J., Pestiaux, P., Zimmerman, H. & Hermelin, O. 1986: Palaeoclimatic and palaeoceanographic development in the Pliocene North Atlantic: *Discoaster accumulation and coarse fraction data. In* Summerhayes, C.P. & Shackleton, N.J. (eds.): *North Atlantic Paleoceanography*, 231–242. *Geological Society Special Publication 21*.

Bailey, J.W. 1851: Microscopical examination of soundings made by the United States Coast Survey of the Atlantic coast of the United States. *Smithsonian Contribution to Knowledge 2*, 1–15.

Bandy, O.L. 1954: Aragonite tests among the Foraminifera. *Journal of Sedimentary Petrology 24*, 60–61.

Bandy, O.L. 1961: Distribution of foraminifera, radiolaria and diatoms in sediments of the Gulf of California. *Micropaleontology 7*, 1–26.

Bandy, O.L. 1963: Larger living foraminifera of the continental borderland of Southern California. *Contributions from the Cushman Foundation for Foraminiferal Research 14*, 121–126.

Bandy, O.L. & Rodolfo, K.S. 1964: Distributions of foraminifera and sediments, Peru–Chile Trench area. *Deep-Sea Research 11*, 817–837.

Bandy, O.L., Ingle, J.C., Jr. & Resig, J.M. 1965: Foraminiferal trends, Hyperion outfall, California. *Limnology and Oceanography 10*, 314–332.

Barker, R.W. 1960: Taxonomic notes on the species figured by H.B. Brady in his report on the foraminifera dredged by H.M.S. Challenger during the years 1873–1876. *Society of Economic Paleontologists and Mineralogists Special Publication 9*, 1–238.

Becker, L.E. & Dusenbury, A.N. Jr. 1958: Mio-Oligocene (Aquitanian) foraminifera from the Goajira Peninsula, Colombia. *Cushman Foundation for Foraminiferal Research, Special Publication 4*, 1–48.

Belanger, P.E. & Streeter, S.S. 1980: Distribution and ecology of benthic foraminifera in the Norwegian–Greenland Sea. *Marine Micropaleontology 5*, 401–428.

Belanger, P.E. & Berggren, W.A. 1986: Neogene benthic foraminifera of the Hatton–Rockall Basin. *Micropaleontology 32*, 324–356.

Belanger, P.E., Curry, W.B. & Matthews, R.K. 1981: Core-top evaluation of benthic foraminiferal isotope ratios for paleoceanographic interpretations. *Palaeogeography Palaeoclimatology Palaeoecology 33*, 205–220.

Belford, D.J. 1966: Miocene and Pliocene smaller foraminifera from Papua and New Guinea. *Department of Natural Development, Bureau of Mineral Resources, Geology, and Geophysics, Bulletin 79*, 1–305.

Berger, W.H. 1973: Deep-sea carbonates: Pleistocene dissolution cycles. *Journal of Foraminiferal Research 3*, 187–195.

Berger, W.H. & Killingley, J.S. 1982: Box cores from the equatorial Pacific: $^{14}$C sedimentation rates and benthic mixing. *Marine Geology 45*, 93–125.

Berger, W.H. & Herguera, J.C. 1992: Reading the sedimentary record of the oceans productivity. *In* Falkowski, P.G. & Woodhead, A.D. (eds.): *Primary Productivity and Biogeochemical Cycles in the Ocean*, 455–486. Plenum, New York, N.Y.

Berger, W.H., Herguera, J.C., Lange, C.B. & Schneider, R. 1994: Paleoproductivity: flux proxies versus nutrient proxies and other problems concerning the Quaternary productivity record. *In* R. Zahn, T.F. Pedersen, M.A. Kaminski & L. Labeyrie (eds.): *Carbon Cycling in the Glacial Ocean: Constraints on the Ocean's Role in the Global Change*, 385–411. Springer, Berlin.

Berggren, W.A., Benson, R.H., Haq, B.U., Riedel, W.R., Sanfilippo, A., Schrader, H.-J. & Tjalsma, R.C. 1976: The El Cuervo Section (Andalusia, Spain). Micropaleontologic anatomy of an early Late Miocene lower bathyal deposit. *Marine Micropaleontology 1*, 195–247.

Berggren, W.A., Kent, D.V. & Van Couvering, J.A. 1985: Neogene geochronology and chronostratigraphy. *Geological Society of London, Memoir 10*, 199–260.

Bermúdez, P.J. 1949: Tertiary smaller foraminifera of the Dominican Republic. *Cushman Laboratory for Foraminiferal Research, Special Publication 24*, 1–322.

Bermúdez, P.J. 1952: Estudio sistematico de los foraminiferos rotaliformes. *Boletin de Geologia 2*, 1–230. [Caracas, Venezuela.]

Bermúdez, P.J. & Seiglie, G.A. 1963: Estudio sistematico de los foraminiferos del Golfo de Cariaco. *Instituto Oceanografico Universidad de Oriente, Boletin 2*, 1–267.

Berthelin, G. 1880: Memoire sur les foraminifères fossils de l'etage Albien de Moncley (Doubs). *Mémoires de la Société Géologique de France, Serie 3, 1*, 1–84.

Bock, W.D. 1970: *Hyalinea baltica* and the Plio-Pleistocene boundary in the Caribbean Sea. *Science 170*, 847–848.

Bock, W.D. 1971: A handbook of the benthonic foraminifera of Florida Bay and adjacent waters. *In* Jones, J.I. & Bock, W.D. (eds.): *A Symposium of Recent South Florida Foraminifera: Miami Geological Society Memoir 1*, 1–72.

Boersma, A. 1984a: Oligocene and other Tertiary benthic foraminifers from a depth traverse down Walvis Ridge, Deep Sea Drilling Project Leg 75, southeast Atlantic. *In* Hay, W.W., Sibuet, J.C. *et al.* (eds.): *Initial Reports of the Deep Sea Drilling Project 75*, 657–669. United States Government Printing Office, Washington, D.C.

Boersma, A. 1984b: Pliocene planktonic and benthic foraminifers from the southeastern Atlantic Angola Margin. Leg 75, Site 532, Deep Sea Drilling Project. *In* Hay, W.W., Sibuet J.C. *et al.* (eds.): *Initial Reports of the Deep Sea Drilling Project 75*, 657–669, United States Government Printing Office, Washington, D.C.

Boersma, A. 1986: Biostratigraphy and biogeography of Tertiary bathyal benthic foraminifers. Tasman Sea, Coral Sea, and on the Chatham Rise (Deep Sea Drilling Project, Leg 90). *In* Kennet, J.P., von der Borch, C.C., *et al.*: *Initial Reports of the Deep Sea Drilling Project 90:2*, 961–1036. United States Government Printing Office, Washington, D.C.

Bolli, H.M., Boudreaux, J.E., Emiliani, C., Hay, W.W., Hurley, R.J. & Jones, J.I. 1968: Biostratigraphy and paleotemperatures of a section cored on the Nicaragua Rise, Caribbean Sea. *Geological Society of America Bulletin 79*, 459–470.

Boltovskoy, E. 1958: Problems in taxonomy and nomenclature exemplified by *Nonion affinis* (Reuss). *Micropaleontology 4*, 193–200.

Boltovskoy, E. 1959: Recent foraminifera of southern Brazil and their relation with those of Argentina and West Indies. *Argentina, Servicio Hidrografia Naval, Publicacion H 1005*, 1–124.

Boltovskoy, E. 1976: Distribution of Recent foraminifera of the South American region. *In* Hedley, R.H. & Adams, C.G. (eds.): *Foraminifera 2*, 717–736. Academic Press, New York, N.Y.

Boltovskoy, E. 1978: Late Cenozoic benthonic foraminifera of the Ninetyeast Ridge (Indian Ocean). *Marine Geology 26,* 139–175.

Boltovskoy, E. 1980: On the benthic bathyal zone foraminifera as stratigraphic guide fossils. *Journal of Foraminiferal Research 10,* 163–172.

Boltovskoy, E. 1981: Benthic late Cenozoic foraminifera of DSDP Site 173 and comparison with the same faunas of other sites. *Revue de Micropaleontologie 23,* 212–217.

Boltovskoy, E. & de Kahn, G.G. 1982: Foraminiferos bentonicos calcareos uniloculares del Cenozoico superios de Atlantico sur. *Revista de la Asociacion Geologica Argentina 37,* 408–448.

Boltovskoy, E. & Totah, V. 1985: Diversity, similarity and dominance in benthic foraminiferal fauna along one transect of the Argentine shelf. *Revue de Micropaleontologie 28,* 23–31.

Boltovskoy, E., Watanabe, S., Totah, V.I. & Vera Ocampo, J. 1992: Cenozoic benthic bathyal foraminifers of DSDP Site 548 (North Atlantic). *Macropaleontology 38,* 183–207.

Boltovskoy, E., Giussani, G. & Wright, R. 1980: *Atlas of Benthic Shelf Foraminifera of the Southwest Atlantic.* 147 pp. Junk, The Hague.

[Boomgaart, L. 1949: Smaller foraminifera from Bodjonegoro (Java). 175 pp. Unpublished Ph.D. Dissertation, University of Utrecht.]

Boyle, E.A. 1990: Quaternary deep water paleoceanography. *Science 249,* 863–870.

Brady, H.B. 1878: On the reticularian and radiolarian Rhizopoda (foraminifera and polycystina) of the North-Polar Expedition of 1875–1876. *Annals and Magazine of Natural History, Series 5, 1,* 425–440.

Brady, H.B. 1881: Notes on some of the reticularian Rhizopoda of the 'Challenger' Expedition. Part III, 1 – Classification, 2 – Further notes on new species, 3 – Note on *Biloculina* mud. *Quaterly Journal of Microscopical Science, New Series, 21,* 31–71.

Brady, H.B. 1884: Report on the foraminifera dredged by HMS Challenger, during the years 1873–1876. *Report of Scientific Results of the Exploration Voyage of HMS Challenger. Zoology 9,* 1–814.

Brady, H.B., Parker, W.K. & Jones, T.R. 1888: On some foraminifera from the Abrohlos Bank. *Transactions of the Zoological Society of London 12:7,* 211–239.

Bremer, M.L., Briskin, M. & Berggren, W.A. 1980: Quantitative paleobathymertry and paleoecology of the Late Pliocene – Early Pleistocene foraminifera of Le Castella (Calabria, Italy). *Journal of Foraminiferal Research 10,* 1–30.

Bremer, M. & Lohmann, G.P. 1982: Evidence of primary control of the distribution of certain Atlantic Ocean benthonic foraminifera by degree of carbonate saturation. *Deep Sea Research 29,* 987–998.

Brotzen, F. 1936: Foraminiferen aus den schwedischen, untersten Senon von Eriksdal in Schonen. *Avhandlingar och Uppsatser från Sveriges Geologiska Undersökning 30 (Ser. C 396),* 1–206.

Brotzen, F. 1942: Die Foraminiferengattung *Gavelinella* nov. gen. und die systematik der Rotaliiformis. *Avhandlingar och Uppsatser från Sveriges Geologiska Undersökning 36 (Serie C 451),* 1–60.

Brotzen, F. 1948: The Swedish Paleocene and its foraminiferal fauna. *Sveriges Geologiska Undersökning Avhandlingar 42 (Serie C 493),* 1–140.

Brunner, C.A. 1984: Evidence for increased volume transport of the Florida Current in the Pliocene and Pleistocene. *Marine Geology 54,* 223–235.

Buchner, P. 1940: Die Lagenen des Golfes von Neapel und der marinen Ablagerungen auf Ischia. *Nova Acta Leopoldiana, New Series, 9,* 363–560.

Burke, S.C. 1981: Recent benthic foraminifera of the Ontong–Java Plateau. *Journal of Foraminiferal Research 11,* 1–19.

Burke, S.K., Berger, W.H., Coulbourn, W.T. & Vincent, E. 1993: Benthic foraminifera in box core ERDC 112, Ontong Java Plateau. *Journal of Foraminiferal Research 23,* 1939.

Butt, A. 1980: Biostratigraphic and paleoenvironmental analysis of the sediments at the Emperor Seamounts, DSDP Leg 55, northwestern Pacific. Cenozoic foraminifers. *In* Jackson, E.D., Koisumi, I. *et al.* (eds.): *Initial Reports of the Deep Sea Drilling Project 55,* 289–325. United States Government Printing Office, Washington, D.C.

Bykova, N.K. 1959: *In* D.M. Rauzer-Chernovsova & A.V. Fursenko (eds.): *Osnovy Paleontologii. Obshchaya Chast' Prosteyshej.* [*Principles of Paleontology. Part I, Protozoa*]. 368 pp. Akademiya Nauk SSSR.

Caralp, M.H. 1988: Late Glacial to Recent Deep-Sea Benthic Foraminifera from the Northeastern Atlantic (Cadiz Gulf) and Western Mediterranean (Alborean Sea): Paleoceanographic Results. *Marine Micropaleontology 13,* 265–289.

Caprariis, P. de, Lindemann, R.H. & Collins, C.M. 1976: A method of determining optimum sample size in species diversity studies. *Journal of International Association in Mathematical Geology 8,* 575–581.

Carpenter, W.B. Parker, W.K. & Jones T.R. 1862: Introduction to the study of the foraminifera. *Ray Society,* 1–319.

Casey, R.E., McMillen, K.J. & Bauer, M.A. 1975: Evidence for and paleoceanographic significance of relict radiolarian populations in the Gulf of Mexico and Caribbean. *Geological Society of America Abstracts with Programs 7,* 1022–1023.

Chapman, F. 1900: On some new and interesting Foraminifera from Funafuti Atoll, Ellice Islands. *Journal of Linnean Society of London, Zoology 28,* 1–27.

Chapman, F., Parr, W.J. Jr. 1937: Foraminifera: Australasian Antarctic Expedition 1911–1914. *Scientific Results, Series C (Zoology and Botany), 1,* 1–190.

Chapman, F., Parr, W.J. Jr. & Collins, A.D. 1934: Tertiary foraminifera of Victoria, Australia – the Balcombian deposits of Port Phillip, Part III. *Journal of the Linnean Society of London, Zoology 38 [1932–1934],* 553–557.

Coates, A.G., Jackson, J.B.C., Collins, L.S., Cronin, T.M., Dowsett, H.J., Bybell, L.M., Jung, P. & Obando, J.A. 1992: Closure of the Isthmus of Panama: The nearshore marine record of Costa Rica and western Panama. *Geological Society of America Bulletin 104,* 814–828.

Coggi, L. & Alliata, E.D.N. 1950: Pliocene e Pleistocene nel' colle di *S. colombano* al Lambro (Lombardia). *18th International Geological Congress, Great Britain 9,* 19–25.

Cole, F.E. 1981: Taxonomic notes on the bathyal zone benthonic foraminiferal species off northeast Newfoundland. *Bedford Institute of Oceanography, Report Series BI-R-81-7,* 1–121.

Collins, L.S. 1990: The correspondence between water temperature and coiling direction in *Bulimina. Paleoceanography 5,* 289–294.

Colom, G. 1950: Estudio de los Foraminiferos de muestras de fondo recogidas entre los Cabos Juby y Bojador. *Bolletino de Institution Espagnol Ocean 28,* 28–45.

Corliss, B.H. 1978: Studies of deep-sea benthonic foraminifera in the southeast Indian Ocean. *Antarctic Journal 13,* 116–117.

Corliss, B.H. 1979: Taxonomy of Recent deep-sea benthonic foramifera from the southeast Indian Ocean. *Micropaleontology 25,* 1–19.

Corliss, B.H. 1982: Linkage of North Atlantic and Southern Ocean Deepwater circulation during glacial intervals. *Nature 298,* 458–460.

Corliss, B.H. 1983: Distribution of Holocene deep-sea benthonic foraminifera in the southwest Indian Ocean. *Deep Sea Research 30,* 95–117.

Corliss, B.H. 1985: Microhabitats of benthic foraminifera within deep-sea sediments. *Nature 314,* 435–438.

Corliss, B.H. 1991: Morphology and microhabitat preferences of benthic foraminifera from northwest Atlantic Ocean. *Marine Micropaleontology 17,* 195–236.

Corliss, B.H. & Honjo, S. 1981: Dissolution of deep-sea benthonic foraminifera. *Micropaleontology 27,* 356–378.

Corliss, B.H., Martinson, D.G. & Keffer, T. 1986: Late Quaternary deep-ocean circulation. *Geological Society of America Bulletin 97,* 1106–1121.

Corliss, B.H. & Chen, C. 1988: Morphotype patterns of Norwegian Sea deep-sea benthic foraminifera and ecological implications. *Geology 16,* 716–719.

Corliss, B.H. & Emerson, S. 1990: Distribution of Rose Bengal stained deep-sea benthic foraminifera from the Nova Scotian continental margin and Gulf of Maine. *Deep-sea Research 37,* 381–400.

Corliss, B.H. & Fois, E. 1990: Morphotype analysis of deep-sea benthic foraminifera from the northwest Gulf of Mexico. *Palaios 5,* 589–605.

Costa, O.G. 1855: Foraminiferi fossili delle marine Terziarie di Messina. *Memoire della Reale Accademia Scienza Napoli 2,* 128– 147, 367–373.

Culver, S.J. 1988: New foraminiferal depth zonation of the northwestern Gulf of Mexico. *Palaios 3,* 69–85.

Crowley, T.J. 1991: Modeling Pliocene warmth. *Quaternary Scientific Reviews 10*, 275–280.

[Culp, S.K. 1977: Recent benthic foraminifera of the Ontong Java Plateau. 146 pp. M.S. Thesis, University of Hawaii, Honolulu, Hawaii.]

Cushman, J.A. 1911: A monograph of the foraminifera of the North Pacific Ocean, Part 2. Textulariidae. *United States National Museum Bulletin 71:2*, 1–108.

Cushman, J.A. 1913: A monograph of the foraminifera of the North Pacific Ocean, Part III. Lagenidae. *United States National Museum Bulletin 71:3*, 1–125.

Cushman, J.A. 1914: A monograph of the foraminifera of the North Pacific Ocean, Part IV. Chilostomellidae, Globigerinidae, Nummulitidae. *United States National Museum Bulletin 71:4*, 1–46.

Cushman, J.A. 1915: A monograph of the foraminifera of the North Pacific Ocean, Part V. Rotalidae. *United States National Museum Bulletin 71:5*, 1–87.

Cushman, J.A. 1917a: A monograph of the foraminifera of the North Pacific, Part VI. Miliolidae. *Proceedings of the United States National Museum Bulletin 71*, 1–108.

Cushman, J.A. 1917b: New species and varities of foraminifera from the Philippines and adjacent waters. *Proceedings of the United States National Museum 51*, 651–662.

Cushman, J.A. 1921: Foraminifera of the Philippine and adjacent seas-contributions to the biology of the Philippine Archipelago and adjacent regions. *United States National Museum Bulletin 104:3*, 1–149.

Cushman, J.A. 1922: The foraminifera of the Atlantic Ocean, Part 3, Textulariidae. *United States National Museum Bulletin 104:3*, 1–149.

Cushman, J.A. 1923: The foraminifera of the Atlantic Ocean, Part 4 Laginidae. *United States National Museum Bulletin 104:4*, 1–228.

Cushman, J.A. 1924: The foraminifera of the Atlantic Ocean. Part 5. Chilostomellidae and Globigerinidae. *United States National Museum Bulletin 104:5*, 1–55.

Cushman, J.A. 1925a: Foraminifera of tropical Central Pacific. *Bernice Payahi Bishop Museum Bulletin 27, Tanager Expedition Publication 1*, 121–144.

Cushman, J.A. 1925b: Notes on the genus *Cassidulina*. *Contributions from the Cushman Laboratory for Foraminiferal Research 1*, 51–60.

Cushman, J.A. 1926a: The genus *Chilostomella* and related genera. *Contributions from the Cushman Laboratory for Foraminiferal Research 1*, 73–80.

Cushman, J.A. 1926b: Foraminifera of the typical Monterey of California. *Contributions from the Cushman Laboratory of Foraminiferal Research 2:3*, 53–69.

Cushman, J.A. 1927a: An outline of a re-classification of the foraminifera. *Contributions from the Cushman Laboratory for Foraminiferal Research 3:1*, 1–105.

Cushman, J.A. 1927b: New interesting foraminifera from Mexico and Texas *Contributions from the Cushman Laboratory for Foraminiferal Research 3:2*, 111–117.

Cushman, J.A. 1927c: Additional notes on the genus *Pleurostomella*. *Contributions from the Cushman Laboratory for Foraminiferal Research 3*, 156–157.

Cushman, J.A. 1929a: The foraminifera of the Atlantic Ocean, Part 6. Miliolidae, Ophthalmidiidae, and Fischerinidae. *United States National Museum Bulletin 104:6*, 1–129.

Cushman, J.A. 1929b: *Planulina ariminensis* d'Orbigny and *P. wuellerstorfi* (Schwager). *Contributions from the Cushman Laboratory for Foraminiferal Research 5*, 102–105

Cushman, J.A. 1930: The foraminifera of the Atlantic Ocean, Part 7. Nonionidae, Camerinidae, Peneroplidae and Alveolinellidae. *United States National Museun Bulletin 104:7*, 1–79.

Cushman, J.A. 1931: The foraminifera of the Atlantic Ocean. Part 8. Rotaliidae, Amphisteginidae, Calcarinidae, Cymbaloporettidae, Globorotaliidae, Anomalinidae, Planorbulinidae, Rupertiidae, and Homotremidae. *United States National Museum Bulletin 104:8*, 1–179.

Cushman, J.A. 1932a: The foraminifera of tropical Pacific collections of the 'Albatross,' 1899–1900, Part 1. Astrorhizidae to Trochamminidae. *United States National Museum Bulletin 161:1:1*, 1–88.

Cushman, J.A. 1933a: The foraminifera of the tropical Pacific collection of the 'Albatross,' 1899–1900, Part 2. Lagenidae, Alveolinellidae. *United States National Museum Bulletin 161:1:2*, 1–79.

Cushman, J.A. 1933b: Some new Recent foraminifera from the tropical Pacific. *Contributions from the Cushman Laboratory for Foraminiferal Research 9*, 77–95.

Cushman, J.A. 1933c: Foraminifera – their classification and economic use. *Cushman Laboratory for Foraminiferal Research, Special Publication 4*, 1–349.

Cushman J.A. 1934: Smaller foraminifera from Vitilevu, Fiji *In* Ladd, H.S.: Geology of Vitilevu, Fiji, 102–142. *Bernice Payahi Bishop Museum Bulletin 119*.

Cushman J.A. 1936a: New genera and species of the families Verneuilinidae and Valvulinidae and of the subfamily Virgulininae. *Cushman Laboratory for Foraminiferal Research, Special Publication 6*, 1–71.

Cushman J.A. 1936b: Some new species of *Nonion. Contributions from the Cushman Laboratory of Foraminiferal Research 12*, 63–69.

Cushman, J.A. 1937a: A monography of the foraminiferal family Verneuilinidae. *Cushman Laboratory for Foraminiferal Research, Special Publication 7*, 1–157.

Cushman, J.A. 1937b: A monograph of the foraminiferal family Valvulinidae. Cushman Laboratory for Foraminiferal Research, *Special Publication 8*, 1–210.

Cushman, J.A. 1937c: A monograph of the subfamily Virgininae of the foraminiferal family Buliminidae. *Cushman Laboratory for Foraminiferal Research, Special Publication 9*, 1–228.

Cushman, J.A. 1939a: A monograph of the foraminiferal family Nonionidae. *United States Geological Survey Professional Paper 191*, 1–100.

Cushman, J.A. 1939b: Notes on some foraminifera described by Schwager from the Pliocene of Kar Nicobar. *Journal of the Geological Society of Japan 46*, 149–154.

Cushman, J.A. 1942: The foraminifera of the tropical Pacific collections of the 'Albatross', 1899–1900. Part 3. Heterohelicidae, and Buliminidae. *United States National Museum Bulletin 161:1:3*, 1–67.

Cushman, J.A. 1946: The species of foraminifera named and figured by Fichtel & Moll in 1798 and 1803. *Cushman Laboratory for Foraminiferal Research, Special Publication 17*, 3–16

Cushman, J.A. 1947: New species and varieties of foraminifera from off the southeastern coast of the United States. *Contributions from the Cushman Laboratory for Foraminiferal Research 23*, 86–92.

Cushman, J.A. 1948: Arctic foraminifera. *Cushman Foundation for Foraminiferal Research, Special Publication 23*, 1–79.

Cushman, J.A. & Bermúdez, P.J. 1936: New genera and species of foraminifera from the Eocene of Cuba. *Contributions from the Cushman Laboratory for Foraminiferal Research 12*, 27–38.

Cushman J.A. & Edwards, P.G. 1937: *Astrononion*, a new genus of the Foraminifera and its species. *Contributions from the Cushman Laboratory for Foraminiferal Research 13*, 29–36.

Cushman J.A. & Edwards, P.G. 1938: Notes on the Oligocene species of *Uvigerina* and *Angologerina. Contributions from the Cushman Laboratory for Foraminiferal Research 14*, 74–89.

Cushman, J.A. & Gray, H.B. 1946: A foraminiferal family from the Pliocene of Timm's Piont, California. *Cushman Laboratory of Foraminiferal Research, Special Publication 19*, 1–46.

Cushman, J.A. & Harris, R.W. 1927: Notes on the genus *Pleurostomella. Contributions from the Cushman Laboratory for Foraminiferal Research 3*, 128–135.

Cushman, J.A. & Jarvis, P.W. 1930: Miocene foraminifera from Buff Bay, Jamaica. *Journal of Paleontology 4*, 353–368.

Cushman, J.A. & Jarvis, P.W. 1934: Some interesting new uniserial foraminifera from Trinidad. *Contributions from the Cushman Laboratory for Foraminiferal Research 10*, 71–75.

Cushman, J.A. & McCulloch, I. 1942: Some Virgulininae in the collections of the Allan Hancock Foundation. *Allan Hancock Pacific Expeditions 6*, 179–230.

Cushman, J.A. & McCulloch, I. 1948: The species of *Bulimina* and related genera in the collections of the Allan Hancock Foundation. *Allan Hancock Pacific Expeditions 6*, 231–294.

Cushman, J.A. & McCulloch, I. 1950: Some Lagenidae in the collections of the Allan Hancock Foundation. *Allan Hancock Pacific Expeditions 6*, 295–364.

Cushman, J.A. & Ozawa, Y. 1930: A monograph of the foraminiferal family Polymorphinidae, Recent and fossil. *Proceedings of the United States National Museum 77*, 1–185.

Cushman, J.A. & Parker, F.L. 1931: Recent foraminifera from the Atlantic coast of South America. *Proceedings of the United States National Museum 8*, 1–34.

Cushman, J.A. & Parker, F.L. 1947: *Bulimina* and related foraminiferal genera. *United States Geological Survey, Professional Paper 210-D*, 55–176.

Cushman, J.A. & Stainforth, R.M. 1945: The foraminifera of the Cipero Marl Formation of Trinidad, British West Indies. *Cushman Laboratory of Foraminiferal Research, Special Publication 14*, 1–75.

Cushman, J.A. & Stone, B. 1947: An Eocene foraminiferal fauna from the Chira Shale of Peru. *Cushman Laboratory of Foraminiferal Research, Special Publication 20*, 1–27.

Cushman, J.A. & Thomas, N.L. 1930: Common foraminifera of the east Texas Greensands. *Journal of Paleontology 4*, 33–41.

Cushman, J.A. & Todd, R. 1943: The genus *Pullenia* and its species. *Contributions from the Cushman Laboratory for Foraminiferal Research 19*, 1–24.

Cushman, J.A. & Todd, R. 1945: Miocene foraminifera from Buff Bay, Jamaica. *Cushman Laboratory of Foraminiferal Research Special Publication 15*, 1–73.

Cushman, J.A. & Todd, R. 1949: Species of the genera *Allomorphina* and *Quadrimorphina*. *Contributions from the Cushman Laboratory for Foraminiferal Research 25*, 59–72.

Cushman, J.A., Parker, F.L. & Post, R.J. 1954: Recent foraminifera from the Marshall Islands, Bikini and nearby atolls, Part 2. Oceanography (Biologic). *United States Professional Paper 260-H*, 319–384.

Czjzek, J. 1848: Beitrag zur Kenntniss der fossilen Foraminiferen des Wiener Beckens. *Haidinger's Naturwissenschaftliche Abhandlungen 2:1*, 137–150.

Czjzek, J. 1849: Über zwei neue Arten von Foraminiferen aus dem Tegel von Baden und Mollersdorf: *Berichte über die Mitteilungen der Freunde der Naturwissenschaften in Wien 5 (1848–1849)*, 50–51.

David, M., Campiglio, C. & Darling, R. 1974. Progress in R- and Q-mode analysis: Correspondence analysis and its application to the study of geological processes. *Canadian Journal of Earth Science 11*, 131–146.

Defrance, M.J. 1824: *In* de Blaineville, H.M.D.: *Dictionaire des Sciences Naturelles 32*, 1–567. Levrault, Paris.

Delage, Y. & Hérouard, E. 1896: *Traité de zoologie concrète. Tome 1: La Cellule et les Protozoaires*, 1–584. Paris.

deMenocal, P.B., Oppo, D.W., Fairbanks, R.G. & Prell, W.L. 1992: Pleistocene $\delta^{13}C$ variability of North Atlantic intermediate water. *Paleoceanography 7*, 229–250.

Denne, R.A. & Sen Gupta, B.K. 1988: Abundance variations of dominant benthic foraminifera on the northwestern Gulf of Mexico slope: relationship to bathymetry and water mass boundaries. *Bulletin de l'Institut de Giologie du Bassin d'Aquitaine, Bordeaux 44*, 33–43.

Denne, R.A. & Sen Gupta, B.K. 1991: Association of bathyal fora minifera with water masses in the northwestern Gulf of Mexico. *Marine Micropaleontology 17*, 173–193.

Dervieux, E. 1894: Le Nodosarie terziarie del Piemonte. *Bollettino della Società Geologica Italiana 3*, 597–626.

Dervieux, E. 1899: Foraminiferi terziarii del Piemonte e specialmente sul gen. *Polymorphina* d'Orbigny. *Bollettino della Società Geologica Italiana 18*, 76–78.

Dieci, G. 1959: Foraminiferi Tortoniani di Montegibbio e Castelvetro (Appennino Modense). *Paleontographia Italica 54 (n.s. 24)*, 1–113.

Douglas, R.G. 1973: Benthonic foraminiferal biostratigraphy in the central North Pacific, Leg 17, Deep Sea Drilling Project. *In* Winterer, E.L., Ewing, J.I. *et al.* (eds.): *Initial Reports of the Deep Sea Drilling Project*, 607–672. United States Government Printing Office, Washington, D.C.

Douglas, R.G. & Woodruff, F. 1981: Deep-sea benthic foraminifera, *In* Emiliani, C. (ed.): *The Oceanic Lithosphere. The Sea 7*, 1233–1327. Wiley, New York, N.Y.

Eade, J.V. 1967: New Zealand Recent foraminifera of the families Islandiellidae and Cassidulinidae. *New Zealand Journal of Marine and Freshwater Research 1*, 421–454.

Earland, A. 1934: Foraminifera Part III. The Falklands sector of the Antarctic (excluding South Georgia). *Discovery Reports 10*, 1–208.

Echols, R.J. 1971: Distribution of foraminifera in sediments of the Scotia Sea area, Antarctic waters, *In* Reid, J.L. (ed.): Antarctic Oceanology 1: *Antarctic Research Series, American Geophysical Union 15*, 93–168.

Ehrenberg, C.G. 1838: Ueber dem blossen Auge unsichtbare Kalkthierchen und Kieselthierchen als Hauptbestandteile der Kreidegebirge. *Berichte der Königlich-Preussischen Akademie der Wissenschaften zu Berlin 3*, 192–200.

Ehrenberg, C.G. 1839: Ueber die Bildung der Kreidefelsen und des Kreidemergels durch unsichtbare Organismen. *Abhandlungen der Königlich-Preussischen Akademie der Wissenschaften zu Berlin*, 59–147.

Ehrenberg, C.G. 1840: Eine weitere Erläuterung des Organismus mehrerer in Berlin lebend beobachteter Polythalamien der Nordsee. *Berichte der Königlich-Preussischen Akademie der Wissenschaften zu Berlin 23*.

Emiliani, C., Gartner, S. & Lidz, B. 1972: Neogene sedimentation on the Blake Plateau and the emergence of the Central American isthmus. *Palaeogeography Palaeoclimatology Palaeoecology 11*, 1–10.

Feyling-Hanssen, R.W. 1964: Foraminifera in late Quaternary deposits from the Oslofjord area. *Norges Geologiske Undersøkelse 225*, 1–383.

Fichtel, L. von & Moll, J.P.C. von 1798: *Testacea microscopica aliaque minuta ex generibus Argonauta et Nautilus (Microscopische und andere kleine Schalthiere aus den Geschlechtern Argonaute und Schiffer)*. 123 pp. Camesina, Wien. [Reprinted 1803].

Finlay, H.J. 1939: New Zealand foraminifera, key species in stratigraphy, No. 1. *Transactions of Royal Society of New Zealand 68*, 504–543.

Finlay, H.J. 1940: New Zealand foraminifera, key species in stratigraphy, No. 4. *Transactions of Royal Society of New Zealand 69*, 448–472.

Finlay, H.J. 1947: New Zealand foraminifera, key species in stratigraphy, No. 5. *New Zealand Journal of Science and Technology 28*, 259–292.

Fornasini, C. 1900: Intorno ad alcuni esemplari di foraminiferi Adriatici. *Memoire della Reale Accademia delle Scienze dell'Instituto di Bologna, Classe di Scienze Naturali, Serie 5, 8 [1899–1900]*, 357–402.

Franklin, E.S. 1944: Microfauna from the Carapita Formation of Venezuela. *Journal of Paleontology 18*, 301–319.

Gaby, M.L. & Sen Gupta, B.K. 1985: Late Quaternary benthic foraminifera of the Venezuela Basin. *Marine Geology 68*, 125–144.

Galloway, J.J. & Heminway, C.E. 1941: Tertiary foraminifera of Porto Rico. *New York Academic Sciences, Scientific Survey of Porto Rico and the Virgin Islands 3*, 275–491.

Galloway, J.J. & Morrey, M. 1929: A lower Tertiary foraminiferal fauna from Manta, Equador. *Bulletins of American Paleontology 15*, 7–56.

Galloway, J.J. & Wissler, S.G. 1927: Pleistocene foraminifera from the Lomita Quarry, Palos Verdes Hills, California. *Journal of Paleontology 1*, 35–87.

Galluzzo, J.J., Sen Gupta, B.K. & Pujos, M. 1990: Holocene deep-sea foraminifera of the Grenada Basin. *Journal of Foraminiferal Research 20*, 195–211.

Gardner, J.V. 1982: High-resolution carbonate and organic-carbon stratigraphies for the late Neogene and Quaternary from the western Caribbean and eastern equatorial Pacific, *In* Prell, W.L., Gardner, J.V. *et al.*, *Initial Reports of the Deep Sea Drilling Project 68*, 347–364. United States Government Printing Office, Washington, DC.

Gary, A.C. 1985: A preliminary study of the relationship between test morphology and bathymetry in Recent *Bolivina albatrossi* Cushman, northwestern Gulf of Mexico. *Gulf Coast Association of Geological Societies Transactions 35*, 381–386.

Gary, A.C., Healy-Williams, N. & Ehrlich, R. 1989: Water-mass relationships and morphologic variability in the benthic foraminifer *Bolivina albatrossi* Cushman, northern Gulf of Mexico. *Journal of Foraminiferal Research 19*, 210–221.

Gaydyukov, A.A. & Likashina, N.P. 1988: Distribution patterns of present-day benthic foraminifera of the North Atlantic and Norwegian Sea as indicated by factor analysis. *Oceanology 28*, 344–347.

Gooday, A.J. 1988: A response by benthic foraminifera to the deposition of phytodetritus in the deep sea. *Nature 332*, 70–73.

Gooday, A.J. 1993: Deep-sea benthic foraminiferal species which exploit phytodetritus: characteristic features and controls on distribution. *Marine Micropaleontology 22, 187–206.*

Gooday, A.J. & Lambshead, P.J.D. 1989: Influence of seasonally deposited phytodetritus on benthic foraminiferal populations in the bathyal northeast Atlantic: the species response. *Marine Ecology Progress Series 58*, 53–67.

Gooday, A.J. & Turley, C.M. 1990: Responses by benthic organisms to inputs of organic material to the ocean floor: a review. *Philosophical Transactions of the Royal Society of London, A 331*, 119–138.

Gooday, A.J., Levin, L.A., Linke, P. & Heeger, T. 1992: The role of benthic foraminifera in deep-sea food webs and carbon cycling. *In* Rowe, G.T. & Pariente, V. (eds.): *Deep-sea Food Chains and the Global Carbon Cycle*, 63–91. Kluwer, Dordrecht.

Gordon, A. 1966: Caribbean Sea – Oceanography. *In* Fairbridge, R.W. (ed.): *The Encyclopedia of Oceanography*, 175–181. Reinhold Publication Corporation, New York.

Graham, D.W., Corliss, B.H., Bender, M.L. & Keigwin, L.D., Jr. 1981: Carbon and oxygen isotope disequilibria of Recent deep-sea benthic foraminifera. *Marine Micropaleontology 6*, 483–497.

Grassle, J.F. & Morse-Porteous, L.S. 1987: Macrofaunal colonization of disturbed deep-sea environments and the stucture of deep-sea benthic communities. *Deep-sea Research 34*, 1911–1950.

Grossman, E.L. 1987: Stable isotopes in modern benthic foraminifera: a study of vital effect. *Journal of Foraminiferal Research 17*, 48–61.

Guembel, C.W. 1868: Beitrage zur Foraminiferen fauna der nordalpinen Eozängebilde. *Abhandlungen der Mathematisch- Physikalischen Klasse der Kaiserlich-Bayerischen Akademie der Wissenschaften 10*, [1870], pt. 2, 581–730.

Gupta, A.K. & Srinivasan, M.S. 1990: Response of northern Indian Ocean deep-sea benthic foraminifera to global climates during Pliocene–Pleistocene. *Marine Micropaleontology 16, 77–91.*

Hada, Y. 1931: Report of the biological survey of Mutsu Bay. 19. Notes on the Recent foraminifera from Mutsu Bay. *Science Reports of the Tohoku Imperial University, 4th Series (Biology)*, 6, 45–148.

Haddad, G.A., Droxler, A.W. & Mahr, J.A. 1994: Southern source upper water layer flux into the Caribbean during the last 200,000 years. *AGU Spring Meeting, Baltimore (Abstract)*, 54.

Haeckel, E. 1894: *Systematische Phylogenie. Entwürf eines naturlichen Systems der Organismen auf Grund ihrer Stammesgeschichte. Teil I: Systematische Phylogenie der Protisten und Pflanzen.* 400 pp. Reimer, Berlin.

Hansen, H.J. & Lykke-Andersen, A. 1976: Wall structure and classification of fossil and recent elphidiid and nonionid foraminifera. *Fossils and Strata 10*, 1–37.

Hantken, M. von. 1875: Die Fauna der *Clavulina Szaboi* Schichten; Teil 1 – Foraminiferen. *Jahrbuch der Kaiserlich-Ungarischen Geologischen Anstalt 4*, 1–93.

Herb, R. 1971: Distribution of Recent benthonic foraminifera in the Drake Passage, *In* Llano, G.A. & Wallen, I.E. (eds.): *Biology of the Antarctic Seas IV*, 251–300. *Antarctic Research Series, American Geophysical Union 17.*

Herguera, J.C. 1992: Deep-sea benthic foraminifera and biogenic opal: Glacial to postglacial productivity changes in the the western equatorial Pacific. *Marine Micropaleontology 19*, 79–88.

Herguera, J.C. 1994: Nutrient, mixing and export indices: a 250 kyr productivity record from the western equatorial Pacific. *In* Zahn, R., Pedersen, T.F., Kaminski, M.A. & Labeyrie, L. (eds.): *Carbon Cycling in the Glacial Ocean: Constraints on the Ocean's Role in Global Change*, 481–520. Springer, Berlin.

Herguera, J.C. & Berger, W.H. 1991: Paleoproductivity from benthic foraminifera abundance: glacial and postglacial change in the west-equatorial Pacific. *Geology 19*, 1173–1176.

Hermelin, J.O.R. 1986: Pliocene benthic foraminifera from the Blake Plateau: Faunal assemblages and paleocirculation. *Marine Micropaleontology 10*, 343–370.

Hermelin, J.O.R. 1989: Pliocene benthic foraminifera from the Ontong–Java Plateau (western equatorial Pacific Ocean). Faunal response to the changing paleoenvironment. *Cushman Foundation for Foraminiferal Research, Special Publication 26*, 3–143.

Hermelin, J.O.R. 1991: *Hyalinea balthica* (Schroeter) in lower Pliocene sediments of the northwest Arabian Sea. *Journal of Foraminiferal Research 21*, 244–251.

Hermelin, J.O.R. & Scott, D.B. 1985: Recent benthic foraminifera from the central North Atlantic. *Micropaleontology 31*, 199–220.

Hermelin, J.O.R. & Shimmield, G.B. 1990: The importance of the oxygen minimum zone and sediment geochemistry in the distribution of Recent benthic foraminifera in the northwest Indian Ocean. *Marine Geology 91*, 1–29.

Heron-Allen, E. & Earland, A. 1932: Foraminifera, Part 1. The ice-free area of Falkland Islands and adjacent seas. *Discovery Reports 4*, 291–460.

Hodell, D.A., Kennett, J.P. & Leonard, K.A. 1983. Changes in vertical water mass structure of the Vema Channel during the Pliocene: evidence from the Deep Sea Drilling Project Holes 516A, 517, 518. *In:* Barker, P.F., Carlsson, R.L., Johnson, D.A. *et al., Initial Reports of the Deep Sea Drilling Project 72*, 907– 919. United States Government Printing Office, Washington, D.C.

Hodell, D.A., Williams, D.F. & Kennett, J.P. 1985: Late Pliocene reorganization of deep vertical water-mass structure in the western South Atlantic: Faunal and isotopic evidence. *Geological Society of America Bulletin 96*, 495–503.

Hodell, D.A. & Venz, K. 1992: Toward a high-resolution stable isotopic record of the Southern Ocean during the Pliocene–Pleistocene (4.8 to 0.8 Ma). *The Antarctic Paleoenvironment: A Perspective on Global Change Antarctic Research Series 56*, 265–310.

Hofker, J. 1951: The foraminifera of Siboga Expedition, part 3. Siboga-Expeditie 1899–1900, *Monograph 4a.* 513 pp. Brill, Leiden.

Hofker, J. 1954: Über die Familie Epistomariidae (Foram). *Palaeontographica A 105*, 166–206.

Hurlbert, S.H. 1971: The nonconcept of species diversity: a critique and alternative parameters. *Ecology 52*, 577–586.

Husezia, R. & Maruhasi, M. 1944. A new genus and thirteen new species of foraminifera from the core-sample of Kasiwazaki oil-field, Nigata-ken. Sigenkagaku Kenkyusyo. *Journal (Research Institute for Natural Resources) 1*, 391–400.

Höglund, H. 1947. Foraminifera in the Gullmar Fjord and Skagerak. *Zoologiska Bidrag från Uppsala Universitet 276*, 1–328.

Ilacqua, M.G. 1956: I foraminiferi Calabriani di Rometta superiore (Messina). *Revesta Italiana di Paleontologia 62*, 225–238.

Ingle, J.C. 1973: Neogene foraminifera from the northeastern Pacific Ocean, Leg 18, Deep Sea Drilling Project, *In* Kulm, L.D., von Huene, R. *et al.: Initial Reports of the Deep Sea Drilling Project 18*, 517–567. United States Government Printing Office, Washington, D.C.

Ingle, J.C., Jr. 1980: Cenozoic paleobathymetry and depositional history of selected sequences within the southern California continental borderland. *In* Sliter, W.V. (ed.): *Studies in Marine Micropaleontology and Paleoecology, A Memorial Volume to Orville L. Bandy. Cushman Foundation for Foraminiferal Research, Special Publication 19*, 163–195.

Ingle, J.C., Jr. & Keller, G. 1980: Benthic foraminiferal biofacies of western Pacific margin between 40°S and 32°N, *In* Field, M., Douglas, R.G., Bouma, A.H. *et al.* (eds.): Quaternary depositional enviroments of the Pacific coast. Pacific coast paleogeography Symposium 4. *Pacific Section of the Society of Economic Paleontologists and Mineralogists*, 341–355.

Ingle, J.C., Jr., Keller, G. & Kolpack, R.L. 1980: Benthic foraminiferal biofacies, sediment and water masses of the southern Peru–Chile Trench area, southeastern Pacific Ocean. *Micropaleontology 26*, 113–150.

Jackson, J.B.C., Jung, P., Coates, A.G. & Collins, L.S. 1993: Diversity and extinction of tropical American mollusks and emergence of the Isthmus of Panama. *Science 260*, 1624–1626.

Jansen, E. & Sjøholm, J. 1991: Reconstruction of glaciation over the past 6 Myr from ice-borne deposits in the Norwegian Sea. *Nature 349*, 600–603.

Jedlitschka, H. 1931: Neue Beobachtungen über *Dentalina verneuilli* (d'Orbigny) und *Nodosaria abyssorum* (Brady). *Firgenwald, Reichenberg, Liberec, Czechoslovakia 4*, 121–127.

Jones, F.W. 1874: On some recent forms of Lagenae from deep-sea soundings in the Java Sea. *Transactions from the Linnean Society of London 30*, 45–69.

Jones, T.R. 1875: *In* Griffith, J.W. & Henfrey, A. (eds.): *The Micrographic Dictionary. Edition 3, 1*, 316–320. Van Voorst, London.

Jumars, P.A. & Wheatcroft, R.A. 1989: Responses of benthos to changing food quality and quantity, with a focus on deposit feeding and bioturbation. *In* Berger, W.H., Smetacek, V.S. & Wefer, G. (eds.): *Productivity of the Oceans: Present and Past,* 235–253. Wiley, Chichester.

Joyce, J.E. & Williams, D.F. 1986: Mid-water history of the Gulf of Mexico during the last 130,000 years: benthic foraminiferal and isotopic evidence. *Marine Micropaleontology 10*, 23–34.

Kaiho, K. 1994: Benthic foraminiferal dissolved oxygen index and dissolved oxygen levels in the modern ocean. *Geology 22*, 719–722.

[Kaneps, A.G. 1970: Late Neogene biostratigraphy (planktonic foraminifera), biogeography, and depositional history. 185 pp. Unpublished thesis. Colombia University, New York, N.Y.]

Karrer, F. 1868: Die Miocene Foraminiferenfauna von Kostej im Batat. *Sizungsberichte der Mathematisch-Naturwissenschaftlichen Klasse der Kaiserlichen Akademie der Wissenschaften zu Wien 58*, 121–193.

Karrer, F. 1877: Geologie der Kaiser Franz Josefs Hochquellen Wasserleitung; eine Studie in den Tertiär-Bildungen am Westrande des alpinen Theiles der Niederung von Wiens. *Kaiserlich-Königliche Geologische Reichsanstalt, Abhandlungen 9*, 1–420.

Kawai, K., Uchio, T. Ueno, M. & Hozuki, M. 1950: Natural gas in the vicinity of Otaki, Chiba-ken. *Journal from the Association of Petroleum Geologists 15*, 151–219. [Japanese with English summary.]

Keigwin, L.D., Jr. 1976: Late Cenozoic planktonic foraminiferal biostratigraphy and paleoceanography of the Panama Basin. *Micropaleontology 22*, 419–422.

Keigwin, L.D., Jr. 1978: Pliocene closing of the Isthmus of Panama, based on biostratigraphic evidence from nearby Pacific Ocean and Caribbean Sea cores. *Geology 6*, 630–634.

Keigwin, L.D., Jr. 1982a: Neogene planktonic foraminifers from Deep Sea Drilling Project Sites 502 and 503, *In* Prell, W.L., Gardner, J.V. *et al.: Initial Reports of the Deep Sea Drilling Project 68*, 269–277. United States Government Printing Office, Washington, DC.

Keigwin, L.D., Jr. 1982b: Stable isotope stratigraphy and paleoceanography of sites 502 and 503, *In* Prell, W.L., Gardner, J.V. *et al.: Initial Reports of the Deep Sea Drilling Project 68*, 311– 323. United States Government Printing Office, Washington, DC.

Keigwin, L.D., Jr. 1982c: Isotopic paleoceanography of the Caribbean and East Pacific: Role of Panama uplift in Late Neogene time. *Science 217*, 350–353.

Keigwin, L.D., Jr. 1987: Pliocene stable-isotope record of Deep Sea Drilling Project Site 606: sequential events of [18]O enrichment beginning at 3.1 Ma. *In* Ruddiman, W.F., Kidd, R.B., Thomas, E. *et al.: Initial Reports of the Deep Sea Drilling Project 94 [Leg 94]*, 911–920. United States Government Printing Office, Washington, DC.

Keller, G. 1980: Benthic foraminifers and paleobathymetry of the Japan Trench area, Deep Sea Drilling Project, *In* Scientific Party, *Initial Reports of the Deep Sea Drilling Project 90:2 [Leg 56–57]*, 835–869. United States Government Printing Office, Washington, D.C.

Kempton, R.A. 1979: The structure of species abundance and measurement of diversity. *Biometrics 35*, 307–321.

Kennett, J.P. 1982: *Marine Geology.* 813 pp. Prentice-Hall, Englewood Cliffs, N.J.

Kent, D.V. & Spariosu, D.J. 1982: Magnetostratigraphy of Caribbean Site 502 Hydraulic Piston Cores, *In* Prell, W.L., Gardner,J.V. *et al.: Initial Reports of the Deep Sea Drilling Project 68*, 419–433. United States Government Printing Office, Washington, DC.

Knowlton, N., Weigt, L.A., Solórzano, L.A., Mills, D.K. & Bermingham, E. 1993: Divergence in proteins, mitochondrial DNA, and reproduc-

tive compatibility across the Isthmus of Panama. *Science 260*, 1629–1632.

Kohl, B. 1985: Early Pliocene benthic foraminifera from the Salina Basin, southeastern Mexico. *Bulletins of American Paleontology 88*, 1–173.

Kurihara, K. & Kennett, J.P. 1986: Neogene benthic foraminifers: distribution in depth traverse, southwest Pacific, *In* Kennett, J.P., von der Borch, C.C. *et al.* (eds.): *Initial Reports of the Deep Sea Drilling Project 90:2 [Leg 90]*, 1037– 1077. United States Government Printing Office, Washington, D.C.

Kurihara, K. & Kennett, J.P. 1988: Bathymetric migration of deep-sea benthic foraminifera in the southwest Pacific during the Neogene. *Journal of Foraminiferal Research 18*, 75–83.

Lamarck, J.P.B.A. de M. de. 1804: Suite des mémoires sur les fossiles des environs de Paris. *Annales Museum National d'Historie Naturelle de Paris 5*, 179–188.

Lamarck, J.P.B.A. de M. de. 1812: *Extrait du cours de zoologie du Museum d'Historie Naturelle sur les animaux invertebres.* 127 pp. Paris.

Lambhead, P.J.D. & Gooday, A.J. 1990: The impact of seasonally deposited phytodetritus on epifaunal and shallow infaunal benthic foraminiferal populations in the bathyal northeast Atlantic: the assemblage response. *Deep-sea Research 37*, 1263–1283.

Le, J. & Shackleton, N.J. 1992: Carbonate dissolution fluctuations in the western Equatorial Pacific during the late Quaternary. *Paleoceanography 7*, 21–42.

LeCalvez, Y. 1974: Revision des foraminifères de la collection d'Orbigny. Part 1, Foraminifères de Isles Canaries. *Cahiers de Micropaleontologie 2*, 1–108.

LeCalvez, Y. 1977: Revision des foraminifères de la collection d'Orbigny. Part 2, Foraminifères de l'Ile de Cuba. *Cahiers de Micropaleontologie 2*, 1–127.

Lehman, S.J. & Keigwin, L.D. 1992: Sudden changes in North Atlantic circulation during the last glaciation. *Nature 356*, 757–762.

LeRoy, D.O. & Levinson, S.A. 1974: A deep-water Pleistocene microfossil assemblage from a well in the northern Gulf of Mexico. *Micropaleontology 20*, 1–37.

LeRoy, L.W. 1941a: Small foraminifera from the late Tertiary of the Nederland East Indies. Part 1, Small foraminifera from the late Tertiary of the Sangkoelirang Bay area, East Borneo, Nederland East Indies. *Quarterly Journal from the Colorado School of Mines 36*, 13–62.

LeRoy, L.W. 1941b: Small foraminifera from the late Tertiary of the Nederland East Indies. Part 2, Small foraminifera from the late Tertiary of the Siberoet Island. *Quarterly Journal from the Colorado School of Mines 36*, 63–105.

LeRoy, L.W. 1944: Miocene foraminifera from Sumatra and Java, Nederland East Indies. Part I, Miocene foraminifera of central Sumatra. *Quarterly Journal from the Colorado School of Mines 39*, 7–69.

LeRoy, L.W. 1964: Smaller foraminifera from the Late Tertiary of southern Okinava. *United States Geological Survey Professional Paper 454-F*, 1–58.

Linke, P. & Lutze, G.F. 1993: Microhabitat preferences of benthic foraminifera: a static concept or a dynamic adaptation to optimize food acquisition? *Marine Micropaleontology 20*, 215–233.

Linsley, B.K., Thunell, R.C., Morgan, C. & Williams, D.F. 1985: Oxygen minimum expension in the Sulu Sea, western equatorial Pacific, during the last glacial low stand of sea level. *Marine Micropaleontology 9*, 395–418.

Loeblich, A.R. Jr. & Tappan, H. 1953: Studies of Arctic foraminifera. *Smithsonian Miscellaneous Collections 121*, 1–150.

Loeblich A.R. & Tappan, H. 1961: Remarks on the systematics of the *Sarcodina* (Protozoa), renamed hononyms, and new and validated genera. *Proceedings of the Biological Society of Washington 74*, 213–234.

Loeblich, A.R. Jr. & Tappan, H. 1963: Four new Recent genera of foraminifera. *Journal of Protozoology 10*, 212–215.

Loeblich, A.R. Jr. & Tappan, H. 1964: *Sarcodina – Chiefly 'Thecamoebians' and Foraminiferida. Treatise on Invertebrate Paleontology C, Protista 2 (2 vols.).* 900 pp. Geological Society of America and University of Kansas Press, Lawrence, Kansas.

Loeblich, A.R. Jr. & Tappan, H. 1988: *Foraminiferal Genera and Thier Classification*, 970 pp. Van Nostrand Reinhold, New York.

Lohmann, G.P. 1978: Abyssal benthonic foraminifera as hydrographic indicators in the western South Atlantic Ocean. *Journal of Foraminiferal Research 8*, 6–34.

Lohmann, G.P. 1981. Modern benthic foraminiferal biofacies: Rio Grande Rise [abstract]. *Transactions of the American Geophysical Union 62*, 903.

Loubere, P. 1991: Deep-sea benthic foraminiferal assemblage response to a surface ocean productivity gradient: a test. *Paleoceanography 6*, 193–204.

Loubere, P. & Banonis, G. 1987: Benthic foraminiferal assemblage response to the onset of northern hemisphere glaciation: paleoenvironmental changes and species trends in the northern Atlantic. *Marine Micropaleontology 12*, 161–181.

Lukashina, N.P. 1988: Distribution pattern of benthic foraminifera in the North Atlantic. *Oceanology 28*, 492–497.

Lutze, G.F. & Thiel, H. 1989: Epibenthic foraminifera from elevated microhabitats: *Cibicidoides wuellerstorfi* and *Planulina ariminensis*. *Journal of Foraminiferal Research 19*, 153–158.

Lutze, G.F. & Coulbourne, W.T. 1984: Recent benthic foraminifera from the continental margin of northwest Africa: community structure and distribution. *Marine Micropaleontology 8*, 361–401.

Mackensen, A. 1992: Neogene benthic foraminifers from the southern Indian Ocean (Kerguelen Plateau): biostratigraphy and paleoecology. *Proceedings of the Ocean Drilling Program, Scientific Results 120*, 649–673.

Mackensen, A. & Douglas, R.G. 1989: Down-core distribution of live and dead deep-water benthic foraminifera in box cores from the Weddell Sea and the California continental borderland. *Deep-sea Research 36*, 879–900.

Mackensen, A., Sejrup, H.P. & Jansen, E. 1985: The distribution of living benthic foraminifera on the continental slope and rise off southwestern Norway. *Marine Micropaleontology 9*, 275–306.

Mackensen, A., Grobe, H., Kuhn, G. & Fütterer, D.K. 1990: Benthic foraminiferal assemblages from the eastern Weddell Sea between 68 and 73°S: Distribution, ecology and fossilization potential. *Marine Micropaleontology 16*, 241–283.

Mackensen, A., Fütterer D.K., Grobe, H. & Schmiedl, G. 1993: Benthic foraminiferal assemblages from the eastern South Atlantic Polar Front region between 35° and 57°S: Distribution, ecology and fossilization potential. *Marine Micropaleontology 22*, 33–69.

McCorkle, D.C., Keigwin, L.D., Corliss, B.H. & Emerson, S.R. 1990: The influence of microhabitats on the carbon isotopic composition of deep-sea benthic foraminifera. *Paleoceanography 5*, 161–185.

Malmgren, B.A. 1983: Ranking of dissolution susceptibility of planktonic foraminifera at high latitudes of the south Atlantic Ocean. *Marine Micropaleontology 8*, 183–191.

Malmgren, B.A. & Sigaroodi, M.M. 1985: Standardization of species counts – the usefulness of Hurlbert's diversity-index in paleontology. *Bulletin of the Geological Institutions of the University of Uppsala 10*, 111–114.

Marie, P. 1941: Les Foraminifères de la Craie à *Belemnitella mucronata* du Bassin de Paris. *Memoires de la Museum National d'Historie Naturelle, New Series, 12*, 1–296.

Marks, P. Jr. 1951: A revision of the smaller foraminifera from the Miocene of the Vienna Basin. *Contributions from the Cushman Foundation for Foraminiferal Research 2*, 33–73.

Matoba, Y. & Yamaguchi, A. 1982: Late Pliocene-to-Holocene benthic foraminifers of the Guaymas Basin, Gulf of California. Sites 477 through 481 *In* Curray, J.R, & Moore, D.G. *et al.* (eds.): *Initial Reports of the Deep Sea Drilling Project 64, [Leg. 64]*, 1027–1056. United States Goverment Printning Office, Washington, D.C.

Matsunaga, T. 1963: Benthonic smaller foraminifera from the oil fields of northern Japan. *Science Reports of the Tohoku University, 2nd Series (Geology), 35*, 1–122.

Matthes, H.W. 1939: Die lagenen des deutschen Tertiärs. *Palaeontographica 90*, 49–108.

McCulloch, I. 1977: Qualitative observations on Recent foraminiferal tests with emphasis on the eastern Pacific. *University of Southern California, Publication.* 1078 pp.

McCulloch, I. 1981: Qualitative observations on Recent foraminiferal tests with emphasis on the eastern Pacific. *University of Southern California, Publication 4.* 363 pp.

McDougall, K. 1985: Miocene to Pleistocene benthic foraminifers and paleoceanography of the middle America slope, Deep Sea Drilling Project Leg 84, *In* von Huene, R., Aubouin, J. *et al.* (eds.): *Initial Reports of the Deep Sea Drilling Project 84 [Leg. 84]*, 363–418. United States Goverment Printing Office, Washington, D.C.

McKnight, W.M. 1962: The distribution of foraminifera off parts of the Antarctic coast. *Bulletins of American Paleontology 44*, 65–158.

Mead, G.A. 1985: Recent benthic foraminifera in the Polar Front region of the southwest Atlantic. *Micropaleontology 31*, 221–248.

Mead, G.A. & Kennett, J.P. 1987: The distribution of recent benthic foraminifera in the polar front region, southwest Atlantic. *Marine Micropaleontology 11*, 343–360.

Metcalf, W.G. 1976: Caribbean–Atlantic water exchange through the Anegada–Jungfern Passage. *Journal of Geophysical Research 81*, 6401–6409.

Miao, Q. & Thunell, R.C. 1993: Recent deep-sea benthic foraminiferal distributions in the South China and Sulu Seas. *Marine Micropaleontology 22*, 1–32.

Miao, Q. & Thunell, R.C. 1996: Late Pleistocene – Holocene distribution of benthic foraminifera in the South China Sea and Sulu Sea: Paleoceanographic implications. *Journal of Foraminiferal Research 26*, 9–23.

Miller, K.G. 1983: Eocene–Oligocene paleoceanography of the deep Bay of Biscay: benthic foraminiferal evidence. *Marine Micropaleontology 7*, 403–440.

Miller, K.G. & Lohmann, G.P. 1982: Environmental distribution of Recent benthic foraminifera on the northeast United States continental slope. *Geological Society of America Bulletin 93*, 200–206.

Millett, F.W. 1901: Reports on the Recent foraminifera of the Malay Archipelago, contained in anchor-mud, collected by M.A. Durrand. *Journal of the Royal Microscopical Society 12*, 619–628.

Montagu, G. 1803: *Testacea Britannica, or Natural History of British Shells, Marine, Land, and Fresh water, Including the Most Minute.* 606 pp. Hollis, Romsey.

Montfort, D. de 1808: *Conchyliologie Systèmatique et Classification Méthodique des Coquilles 1.* 409 pp. Schoell, Paris.

Morkhoven, F.R.C.M. Van 1981: Cosmopolitan Tertiary bathyal benthic foraminifera (abstract with accompanying range charts). *Transactions of the Gulf Coast Association of Geological Societies, Supplement 31*, 1–445

Morkhoven, F.R.C.M. Van, Berggren, W.A. & Edwards, S.A. 1986: Cenozoic cosmopolitan deep-water benthic foraminifera. *Bulletin Centres Recherches Exploration-Prodoction Elf-Aquitaine, Mémoires 11*, 1–412, Pau, France.

Murray, J.W. 1971: *An atlas of British Recent Foraminiferids.* 244 pp. Heinemann, London.

Murray, J.W. 1984: Paleogene and Neogene benthic foraminifers from Rockall Plateau. *In* Roberts, D.G., Schnitker, D. *et al.* (eds.): *Initial Reports of the Deep Sea Drilling Project 81*, 503–534. United States Government Printing Office, Washington, D.C.

Murray, J.W. 1988: Neogene bottom water masses in the NE Atlantic Ocean. *Journal of the Geological Society in London 145*, 125–132.

Murray, J.W. 1991: *Ecology and Paleoecology of Benthic Foraminifera.* 397 pp. Wiley, New York, N.Y.

Nørvang, A. 1958: *Islandiella* n.g. and *Cassidulina* d'Orbigny. *Meddelelser Dansk Naturhistorisk Forening 120*, 25–41.

Oberhänsli, H., Müller-Merz, E. & Oberhänsli, R. 1991: Eocene paleoceanographic evolution at 20°–30°S in the Atlantic Ocean. *Palaeogeography Palaeoclimatology Palaeoecology 83*, 173–216.

Oppo, D.W. & Fairbanks, R.G. 1987: Variability in the deep and intermediate water circulation of the Atlantic Ocean during the past 25,000 years: Northern hemisphere modulation of the Southern Ocean. *Earth and Planetary Science Letters 86*, 1–15.

d'Orbigny, A.D. 1826: Tableau métodique de la classe des céphalopes. *Annales des Sciences Naturelles, Serie 1, 7*, 245–314. Paris.

d'Orbigny, A.D. 1839a: Foraminifères, *In* R. de la Sagra (ed.): *Historire Physique, Politique et Naturelle de L'Ile de Cuba.* 224 pp. Arthus Bertrand, Paris.

d'Orbigny, A.D. 1839b: *Voyage dans l'Amérique Méridionale – Foraminifères 5:5.* 86 pp. Levrault, Strasbourg.

d'Orbigny, A.D. 1839c: Foraminifères des îles Canaries, *In* Barker-Webb, P. & Berthelot, S. (eds.): *Historie Naturelle des Iles Canaries 2*, 119–146. Béthune, Paris.

d'Orbigny, A.D. 1846: *Foraminiferes Fossiles du Bassin Tertiaire de Vienne (Autriche).* 312 pp. Gide, Paris.

Palmer, D.K. & Bermúdez, P.J. 1936: An Oligocene foraminiferal fauna from Cuba. *Memoiras de la Sociedad Cubana de Historia Natural 10*, 273–316.

Parker, F.L. 1952: Foraminiferal species off Portsmouth, New Hampshire. *Bulletin of the Harvard Museum of Comparative Zoology 106*, 391–423.

Parker, F.L. 1954: Distribution of foraminifera in the northeast Gulf of Mexico. *Bulletin of the Harvard Museum of Comparative Zoology 111*, 453–588.

Parker, F.L. 1958: Eastern Mediterranean foraminifera. Sediment cores from the Mediterranean Sea and the Red Sea. *Reports of the Swedish Deep Sea Expedition 1947–1948 8*, 217–283.

Parker, F.L. 1964: Foraminifera from the experimental Mohole drilling near Guadalupe Island, Mexico. *Journal of Paleontology 38*, 617–636.

Parker, F.L. 1973: Late Cenozoic biostratigraphy (planktonic foraminifera) of tropical Atlantic deep-sea sections. *Revista Española de Micropaleontologia 5*, 115–208.

Parker, W.C., Clark, M.W., Wright, R.C. & Clark, R.K. 1984: Population dynamics, Paleogene abyssal benthic foraminifers, eastern South Atlantic, Deep Sea Drilling Project, Leg 73. *In* Hsü, K.J., LaBrecque, J.L. *et al.*: *Initial Reports of the Deep Sea Drilling Project 73*, 481–486. United States Goverment Printing Office, Washington, D.C.

Parker, W.K. & Jones, T.R. 1862: *In* Carpenter, W.B., Parker, W.K. & Jones, T.R. (eds.): Introduction to the study of the foraminifera. *Ray Society Publications*, 1–319. London.

Parker, W.K. & Jones, T.R. 1865: On some foraminifera from the North Atlantic and Arctic Ocean, including Davis Straits and Baffin's Bay. *Philosophical Transactions of the Royal Society of London 155*, 325–441.

Parker, W.K., Jones, T.R. & Brady, H.B. 1871: On the nomenclature of the foraminifera. Part XIV – The species enumerated by d'Orbigny in the 'Annales des Sciences Naturelles' 1826, 7. *Annales and Magazine of Natural History, Series 4, 8*, 1–145, 179, 266–283. London.

Parr, W.J. 1947: The lagenid foraminifera and their relationships. *Royal Society Victoria, Proceedings 58*, 116–130.

Parr, W.J. 1950: Foraminifera. British and New Zealand Antarctic Research Expedition, 1929–1931, *Reports, Series B (Zoology and Botany)*, 5, 233–392.

Pedersen, T.F., Pickering, M., Vogel, J.S., Southon, J. & Nelson, D.E. 1988: The response of benthic foraminifera to productivity cycles in the eastern equatorial Pacific: faunal and geochemical constraints on glacial bottom water oxygen levels. *Paleoceanography 3*, 157–168.

Pflum, C.E. & Frerichs, W.D. 1976: Gulf of Mexico deep-water foraminifera. *Cushman Foundation for Foraminiferal Research, Special Publication 14*, 1–125.

Phleger, F.B. 1951: Ecology of foraminifera, northwest Gulf of Mexico. Part 1, Foraminifera distribution. *Geological Society of America Memoir 46:1*, 1–88.

Phleger, F.B. 1952: Foraminifera distribution in some sediment samples from the Canadian and Greenland Arctic. *Contributions from the Cushman Foundation for Foraminiferal Research 3*, 80–89.

Phleger, F.B. & Parker, F.L. 1951: Ecology of foraminifera, northwest Gulf of Mexico. Part II, Foraminifera species. *Geological Society of America Memoir 46:2*, 1–64.

Phleger, F.B., Parker, F.L. & Pierson, J.F. 1953: North Atlantic foraminifera. *Reports of the Swedish Deep Sea Expedition 7*, 3–122.

Poag, C.W. 1981: *Ecologic Atlas of Benthic Foraminifera of the Gulf of Mexico.* 174 pp. Marine Science International, Woods Hole.

Poag, C.W. 1984: Distribution asnd ecology of deep-water benthic foraminifera in the Gulf of Mexico. *Palaeogeography Palaeoclimatology Palaeoecology 48*, 25–37.

Prell, W.D. 1984: Covarience of patterns of foraminiferal $\delta^{18}O$: An evaluation of Pliocene ice volume changes near 3.2 million years ago. *Science 266*, 692–694.

Prell, W.D., Gardner, J.W. *et al.* 1982: Site 502: Colombia Basin, western Caribbean Sea, *In* W.D. Prell, J.W. Gardner, *et al.*: *Initial Reports of the Deep Sea Drilling Project 68*, 15–162. United States Government Printing Office, Washington, D.C.

Qvale, G. & van Weering, T.C.E. 1985: Relationship of surface sediments and benthic foraminiferal distribution patterns in the the Norwegian Channel (Northern North Sea). *Marine Micropaleontology 9*, 469–488.

Rahaghi, A. 1977: Sur le genre *Cuvillierella* n.gen. et quelques nouvelles espèces des Faluns de Saubrigues (Landes, France). *Revue de Micropaléontologie 19*, 166–171.

Rathburn, A.E. & Corliss, B.H. 1994: The ecology of living (stained) deep-sea benthic foraminifera from the Sulu Sea. *Paleoceanography 9*, 87–150.

Raymo, M.E., Hodell, D.A. & Jansen, E. 1992: Response of deep ocean circulation to initiation of Northern Hemisphere glaciation. *Paleoceanography 7*, 645–672.

Rau, W.W. 1948: Foraminifera from the Porter Shale (Lincoln Formation), Grays Harbor County, Washington. *Journal of Paleontology 22:2*, 152–174.

Reid, J.L., Nowlin, W.D., Jr. & Patzert, W.C. 1976: On the characteristics and circulation of southwestern Atlantic Ocean. *Journal of Physical Oceanography 7*, 62–91.

Renz, H.H. 1948: Stratigraphy and fauna of the Agua Salada group, State of Falcon, Venezuela. *Geological Society of America Memoir 32*, 1–219.

Resig, J.M. 1976: Benthic foraminiferal stratigraphy, eastern margin, Nazca Plate. *In* Yeats, R.S., Hart, R.S. *et al.* (eds.): *Initial Reports of the Deep Sea Drilling Project 34 [Leg. 34]*, 743–759. United States Goverment Printing Office, Washington, D.C.

Resig, J.M. 1981: Biography of benthic foraminifera of the northern Nasca Plate and adjacent continental margin. *In* Kulm, L.D. Dymond, J., Dasch, E.J. & Hussong, D.M. (eds.): Nasca Plate: Chrustal Formation and Andean Convergence: *Geological Survey of America Memoir 154*, 467–507.

Resig, J.M. 1990: Benthic foraminiferal stratigraphy and paleoenvironments off Peru, Leg 112, *In* Suess, E., von Huene, R. *et al.*: *Proceedings of the Ocean Drilling Program, Scientific Results 112* 263–296. College Station, TX (Ocean Drilling Program).

Reuss, A.E. 1845: *Die Versteinerungen der böhmischen Kreideformation 1*, 1–58. Schweizerbart, Stuttgart.

Reuss A.E. 1850: Neue Foraminiferen aus dem Österreichischen Tertiärbecken. *Denkschriften der Mathematisch-Naturwissenschaftlichen Klasse der Kaiserlichen Akademie der Wissenschaften zu Wien 1*, 365–390.

Reuss, A.E. 1851a: Ueber die fossilen Foraminiferen und Entomostraceen der Septarientone der Umgegend von Berlin. *Zeitschrift der Deutschen Geologischen Gesellschaft 3*, 49–91.

Reuss, A.E 1851b: Die Foraminiferen und Entomostraceen des Kreidemergels von Lemberg. *Haidinger's Naturwissenschaftliche Abhandlungen 4*, 17–52.

Reuss, A.E. 1860: Die Foraminiferen der Westphälischen Kreide- Formation. *Sitzungberichte der Mathematisch- Naturwissenschaftlichen Klasse der Kaiserlichen Akademie der Wissenschaften zu Wien 40*, 147–238.

Reuss, A.E. 1863: Die Foraminiferen-Familie der Lagenideen. *Sitzungberichte der Mathematisch-Naturwissenschaftlichen Klasse der Kaiserlichen Akademie der Wissenschaften zu Wien 46:1*, 308–342.

Reuss, A.E. 1870: Die Foraminiferen Septarientones von Pietzpuhl. *Sitzungberichte der Mathematisch-Naturwissenschaftlichen Klasse der Kaiserlichen Akademie der Wissenschaften zu Wien 62:1*, 446–487.

Risso, A. 1826: *Histoire naturelle des principales productions de l'Europe Meridionale et particulierement des celles des environs de Nice et des Alpes maritimes.* 439 pp. Levrault, Paris.

Rodrigues, C.G., Hooper, K. & Jones, P.C. 1980: The apertural structures of *Islandiella* and *Cassidulina*. *Journal of Foraminiferal Reasearch 10*, 48–60.

Rosen, D.E. 1975. A vicariance model of Caribbean biogeography. *Systematic Zoology 24*, 431–464.

Ross, C.R. 1984: *Hyalinea balthica* and its Late Quaternary paleoclimatic implications, Strait of Sicily. *Journal of Foraminiferal Research 14*, 134–139.

Ross, C.R. & Kennett, J.P. 1983: Late Quaternary paleoceanography as recorded by benthonic foraminifera in Strait of Sicily sediment sequences. *Marine Micropaleontology 8*, 315–336.

Ruddiman, W.F., Backman, J., Balhauf, J., Hooper, P., Keigwin, L., Miller, K., Raymo, M. & Thomas, E. 1987: 47. Leg 94 Paleoenvironmental synthesis. *In* Ruddiman, W.R., Kidd, R.B., Thomas, E. *et al.* (eds.): *Initial Reports of the Deep Sea Drilling Project 94 [Leg. 94]*, 1207–1215. United States Goverment Printing Office, Washington, D.C.

Saito, T. 1976: Geological significance of coiling direction in the planktonic foraminifera *Pulleniatina*. *Geology 4*, 305–309.

Schlict, E. von. 1870: *Die Foraminiferen des Septarienthones von Pietzpuhl*. Pls. 1–38. Berlin.

Schlumberger, C. 1887: Note sur le genre *Planispirina*. *Bulletin de la Société Zoologique de France 12*, 105–118.

Schnitker, D. 1971: Distribution of foraminifera on the North Carolina continental shelf. *Tulane Studies in Geology and Paleontology 8*, 169–215.

Schnitker, D. 1974: West Atlantic abyssal circulation during the past 120,000 years. *Nature 248*, 385–387.

Schnitker, D. 1979: The deep waters of the western North Atlantic during the past 24,000 years, and re-initiation of the Western Boundary Undercurrent. *Marine Micropaleontology 4*, 265–280.

Schnitker, D. 1980: Quaternary deep-sea benthic foraminifers and bottom-water masses. *Annual Review of Earth and Planetary Sciences 8*, 343–370.

Schnitker, D. 1994: Deep-sea benthic foraminifers: food and bottom water masses. *In* Zahn, R., Pedersen, T.F., Kaminski, M.A. & Labeyrie, L. (eds.): *Carbon Cycling in the Glacial Ocean: constraints on the Ocean's role in global change*, 539–554. Springer, Berlin.

Schubert, R.J. 1907: Beitrage zu einer naturlichen Systematik der Foraminiferen. *Neues Jahrbuch für Mineralogie, Geologie und Paläontologie, Beilage-Band 25*, 232–260.

Schultze, M.S. 1854: *Ueber der Organismus der Polythalamien (Foraminiferen) nebst Bemerkungen über Rhizopoden in Allgemeinen.* 68 pp. Engelmann, Leipzig.

Schwager, C. 1866: Fossile Foraminiferen von Kar-Nikobar. *Reise der Österreichischen Fregatte Novara um die Erde in Jahren 1857, 1858, 1859, Geologischer Teil, 2:2*, 187–268.

Schwager, C. 1877: Quadro del proposto sistema di classificazione dei foraminiferi con guscio. *Bollettino della Reale Comitato Geologia Italiana 8*, 18–27.

Schwager, C. 1878: Nota su alcuni foraminiferi nuovi del tufo di Stretto presso Girgenti. *Bollettino della Reale Comitato Geologia Italiana 9*, 511–514, 519–529.

Schröder-Adams, C.J., Cole, F.E., Medioli, F.S., Mudie, P.J., Scott, D.B. & Dobbin, L. 1990: Recent Arctic shelf foraminifera: seasonally ice covered vs. perennially ice covered areas. *Journal of Foraminiferal Research 20*, 8–36.

Schroeter, J.S. 1783: *Einleitung in die Conchylien-Kenntnis nach Linné 1.* 860 pp. Gebauer, Halle.

Scott, D.B. & Vilks, G. 1991: Benthonic foraminifera in the surface sediments of the deep-sea Arctic Ocean. *Journal of Foraminiferal Research 21*, 20–38.

Seguenza, G. 1862a: *Dei terreni Terziarii del distretto di Messina. Parte II – Descrizione de foraminiferi monotalamici delle marne miocenische del distretto di Messina*, 1–84. Capra, Messina.

Seguenza, G. 1862b: Prime recherche intorno ai rizopodi fossili delle argille Pleistocenichi dei dintorni di Catania. *Atti della Accademia Gioenia della Scienza Naturali di Catania, Series 2, 18*, 84–126.

Seguenza, G. 1880: Le formazione terziarie nella provincia di Reggio (Calabria). *Memoire della Reale Accademia dei Lincei, Classe di Scienze Fisiche, Matematiche e Naturali, Series 3, 6*, 3–446.

Sen Gupta, B.K. 1988: Water mass relation of the benthic foraminifer *Cibicides wuellerstorfi* in the eastern Caribbean Sea. *Bulletin de l'Institut de Giologie du Bassin d'Aquitaine, Bordeaux, 44*, 23–32.

Sen Gupta, B.K. & Machain-Castillo, M.L. 1993: Benthic foraminifera in oxygen-poor habitats. *Marine Micropaleontology 20*, 183–201.

Shackleton, N.J., Backman, J., Zimmerman, H., Kent, D.V., Hall, M.A., Roberts, D.G., Schnitker, D., Balkauf, J.G., Desprairies, A., Homrighausseen, R., Huddlestun, P., Kenne, J.B., Kaltenback, A.J., Krumsiek, K.A.O., Morton, A.C., Murray, J.W. & Westberg-Smith, J. 1984: Oxygen isotope calibration of the onset of ice-rafting and history o glaciation in the North Atlantic region, *Nature 307*, 620–623.

Silvestri, A. 1896: Foraminiferi Pliocenici della Provincia di Siena, parte 1. *Memorie della Accademia Pontificia dei Nuovi Lincei 12*, 1–204.

Silvestri, A. 1904a: Richerche strutturali su alcune forme dei Trubi dei Bonfornello (Palermo), *Memorie della Accademia Pontificia dei Nuovi Lincei 22*, 235–276.

Silvestri, A. 1904b: Formae nuove o poco conosciute di Protozoi miocenici piemontesi. *Atti della Reale Accademia delle Scienze di Torino 39*, 4–15.

Silvestri, A. 1924: Fauna Paleogenica di Vasciano presso Todi. *Bollettino della Società Geologica Italiana 42:1 [1923]*, 7–29.

Silvestri, O. 1872: Saggio di studi sulla fauna microscopia fossile appartenente al terreno sunappenino italiano. Mem. 1 – Monografia delle Nodosarie. *Bollettino della Accademia Gioenia delle Sienze Naturali di Catania, Series 3:7*, 1–108.

Smart, C., King, S.C., Gooday, A.J., Murray, J.W. & Thomas, E. 1994: A benthic foraminiferal proxy of pulsed organic matter paleofluxes. *Marine Micropaleontology 23*, 89–100.

Smith, P.B. 1963: Recent foraminifera off Central America, quantitative and qualitative studies of the family Bolivinidae. *United States Geological Survey Professinal Paper 429-A*, 1–39.

Smith, P.B. 1964: Ecology of benthonic species. *United States Geological Survey Professinal Paper 429-B*, 1–55.

Sokal, R.R. & Rohlf, F.J. 1969: *Biometry*. 776 pp. Freeman, San Francisco, Cal.

Souaya, F.J. 1965: Miocene foraminifera of the Gulf of Suez region, U.A.R., Part I – Systematics (Astrorhizoidea–Buliminidea). *Micropaleontology 11*, 301–334.

Stainforth, R.M. 1952: Classification of uniserial calcareous foraminifera. *Contributions from the Cushman Foundation for Foraminiferal Research 3*, 6–14.

Stewart, R.E & Stewart, K.C. 1930: Post-Miocene foraminifera from the Ventura Quadrangle, Ventura County, California. *Journal of Paleontology 4*, 60–72.

Streeter, S.S. 1973: Bottom water and benthonic foraminifera in the North Atlantic: Glacial–Interglacial contrasts. *Quaternary Research 3*, 131–141.

Streeter, S.S. & Shackleton, N.J. 1979: Paleocirculation of the deep North Atlantic. A 150,000 year record of benthic foraminifer and oxygen-18. *Science 203*, 168–170.

Streeter, S.S. & Lavery, S.A. 1982: Holocene and last glacial benthic foraminifera from the slope and rise off estern North America. *Geological Society of America Bulletin 93*, 190–199.

Sverdrup, H.U., Johnson, M.W. & Fleming, R.H. 1942: *The Oceans*. Prentice–Hall, Englewood Cliffs, N.J.

[Tappa, K. 1992: The ecology of recent benthic foraminifera from the South China Sea. 90 pp. Master thesis, Duke University, Durham, N.C.]

Taylor, H., Patterson, R.T. & Choi, H.-W. 1985: Occurrence and reliability of internal morphologic features in some Glandulinidae (Foraminiferida). *Journal of Foraminiferal Research 15*, 18–23.

Thalmann, H.E. 1937: Mitteilungen über Foraminiferer Part III. *Ecologae Geologicae Helvetiae 3*, 337–356.

Thalmann, H.E. 1939: Bibliography and index to new genera, species and varieties of foraminifera for the year 1936. *Journal of Paleontology 21*, 355–395.

Thalmann, H.E. 1942: Nomina Bradyana mutata. *The American Midland Naturalist 28*, 463–464.

Thalmann, H.E. 1950: New names and homonyms in Foraminifera. *Contributions from the Cushman Foundation for Foraminiferal Research 1*, 41–45.

Thiede, J. 1973: Planktonic foraminifera in hemipelagic sediments: shell preservation off Portugal and Morocco. *Geological Society of America Bulletin 84*, 2749–2754.

Thomas, E. 1985: Late Eocene to Recent deep-sea benthic foraminifers from the central equatorial Pacific Ocean. *In* Mayer, L., Theyer, F. *et al.* (eds.): *Initial Reports of the Deep Sea Drilling Project 85*, 655–694. United States Government Printing Office, Washington, D.C.

Thomas, E. 1992: Cenozoic deep-sea circulation: evidence from deep-sea benthic foraminifera. *In: The Antarctic Paleoenvironment: A Perspective on Global Change*, 141–165. *Antarctic Research Series 56*.

Thomas, E. & Vincent, E. 1987: Major changes in benthic foraminifera in the equatorial Pacific before the middle Miocene polar cooling. *Geology 15*, 1035–1039.

Thomas, E. & Vincent, E. 1988: Early to middle Miocene deep-sea benthic foraminifera in the equatorial Pacific. *Revue de Paleobiologie, Special Volume 2 (BENTHOS'86)*, 583–588.

Thomas, F.C., Medioli, F.S. & Scott, D.B. 1990: Holocene and latest Wisconsinan benthic foraminiferal assemblages and paleocirculation history, lower Scotian slope and rise. *Journal of Foraminiferal Research 20*, 212–245.

Thompson, P.R. 1980: Foraminifers from the Deep Sea Drilling Project Sites 434, 435, and 436, Japan Trench. *In:* Scientific Party, *Initial Reports of the Deep Sea Drilling Project 56/57:2*, 775–807. United States Governement Printing Office, Washington, D.C.

Thunell, R.C. 1976: Optimum indices of calcium carbonate dissolution in deep-sea sediments. *Geology 4*, 525–528.

Thunell, R.C., Miao, Q., Calvert, S.E. & Pedersen, T.F. 1992: Glacial–Holocene biogenic sedimentation patterns in the South China Sea: Productivity variations and surface water $pCO_2$. *Paleoceanography 7*, 143–162.

Tjalsma, R.C. & Lohmann, G.P. 1983: Paleocene–Eocene bathyal and abyssal benthic foraminifera from the Atlantic Ocean. *Micropaleontology Special Publication*, 1–90.

Todd, R. 1965: The foraminifera of the tropical Pacific collections of 'Albatross', 1899–1900, Part 4. Rotaliiform families and planktonic families. *United States National Museum Bulletin 161*, 1–139.

Todd, R. 1966: Smaller foraminifera from Guam. *United States Geological Survey Professional Paper 403-I*, 1–41.

Todd, R & Brönnimann, P. 1957: Recent foraminifera and thecamoebina from the eastern Gulf of Paria. *Cushman Foundation for Foraminiferal Research, Special Publication 3*, 1–43.

Todd, R. & Low, D. 1976: Smaller foraminifera from wells on Puerto Rico and St. Croix. *United States Geological Survey Professional Paper 863*, 1–32.

Trauth, F. 1918: Das Eozänvorkommen bei Radstadt im Ponqua und seine Bezeihungen zu den gleichalterigen Ablagerungen bei Kirchberg am Wechsel und Wimpassing am Leithagebirge. *Kaiserliche Akademie der Wissenschaften zu Wien, Mathematisch-Naturwissenschaftliche Klasse 95*, 235.

Tyler, P.A. 1988: Seasonality in the deep sea. *In* Barnes, H. & Barnes, M. (eds.): *Oceanography and Marine Biology*, 227–258. Aberdeen University Press, Aberdeen.

Uchio, T. 1960: Ecology of living benthonic foraminifera from the San Diego, California area. *Cushman Foundation for Foraminiferal Research Special Publication 5*, 1–72.

Ujiié, H. 1990: Bathyal benthic Foraminifera in a piston core from east off the Miyako Islands, Ryukyu Island Arc. *Bulletin of the College of Science, University of the Ryukyus 49*, 1–60.

Vermeij, G.J. 1993: The biological history of a seaway. *Science 260*, 1603–1604.

Vilks, G. 1969: Recent foraminifera in the Canadian Arctic. *Micropaleontology 15*, 35–60.

Voloshinova, N.A. 1958: O novoy sistematike Nonioinid. [On a new systematics of the Nonionidae.] *Microfauna SSSR, Sbornik 9, VNIGRI, 115*, 117–223. (In Russian.)

Voorthuysen, J.H. van. 1950: The quantitative distribution of the Pleistocene, Pliocene and Miocene Foraminifera of boring Zaandam (Netherlands). *Mededelingen van de Geologische Sticht, Nieuwe Serie, 4*, 51–72.

Voorthuysen, J.H. van 1952: A new name for a Pleistocene foraminifer from the Netherlands. *Journal of Paleontology 26*, 680– 681.

[Walch, C. 1978: Recent abyssal benthic foraminifera from the eastern equatorial Pacific. 117 pp. Unpublished M.S. Thesis, University of Southern California, Los Angeles.]

Walker, G. & Boys, W. 1784: Testacea Minuta Rariora, nuperrinae detecta in arena littoris Sandvicensis a Gul. Boys, arm S.A.S. multa addidit, et omnium figuras ope microscopii amplitatas accurate delineavit Geo. Walker. [A collection of the minute and rare shells lately discovered in the sand of seashore near Sandwich by William Boys.] 25 pp. March, London.

Walker, G. & Jacob, E. 1798: *In* Kammacher, F.: *Adam's Essays on the Microscope. Edition 2, With Considerable Additions and Improvements.* 712 pp. Dillon & Keating, London.

Wedekind, P.R. 1937: *Einführung in die Grundlagen der historischen Geologie, Band II. Mikrobiostratigraphie die Korallen- und Foraminiferenzeit.* 136 pp. Stuttgart.

Weinberg, J.R. 1991: Rates of movement and sedimentary traces of deep-sea foraminifera and mollusca in the laboratory. *Journal of Foraminiferal Research 21*, 213–217.

Weston, J.F. & Murray, J.W. 1984: Benthic foraminifera as deep-sea water-mass indicators. *In* Oertli, H.J. (ed): *Benthos'83; 2nd International Symposium in Benthic Foraminifera (Pau, April, 1983)*, 605–610. Elf Aquitaine, Esso REP and Total CFP, Pau and Bordeaux.

Wiesner, H. 1931: Die Foraminiferen. *In* E. von Drygalski: *Deutsche südpolar-Expedition 1901–1903, 20. Zoologie 12*, 53–165.

Williamson, W.C. 1848: On the Recent British species of the genus *Lagena. Annals and Magazine of Natural History, Series 2, 1*, 1–20.

Williamson, W.C. 1858: On the Recent foraminifera of Great Britain. *Ray Society*, 1–107. London.

Whitmore, F.C.J. & Stewart, R.H. 1965: Miocene mammals and Central American seaways. *Science 148*, 180–182.

Woodring, W.P. 1966: The Panama land bridge as a sea barrier. *American Philosophical Society Proceedings 110:6*, 425–433.

Woodruff, F. 1985: Changes in Miocene deep sea benthic foraminiferal distribution in the Pacific Ocean: Relationship to paleoceanography. *Memoir of Geological Society of America 163*, 131–176.

Woodruff, F. & Douglas, R.G. 1981: Response of deep sea benthic foraminifera to Miocene paleoclimatic events, DSDP Site 289. *Marine Micropaleontology 6*, 617–632.

Woodruff, F. & Savin, S.M. 1989: Miocene deep-water oceanography. *Paleoceanography 4*, 87–140.

Worthington, L.W. 1970: The Norwegian Sea as a Mediterranean basin. *Deep Sea Research 17*, 77–84.

Wright, R. 1978: Neogene benthic foraminifers from DSDP Leg 42A, Mediterranean Sea. *In* Hsü, K.J., Montadert, L. *et al. Initial Reports of the Deep Sea Drilling Project 42*, 709–726. United States Goverment Printing Office, Washington, D.C.

Wüst, G. 1964: *Stratification and Circulation in the Antillean–Caribbean Basins, Part 1: Spreading and Mixing of the Water Types, with an Oceanographic Atlas.* 130 pp. Columbia University Press, New York, N.Y.

Zahn, R., Winn, K. & Sarnthein, M. 1986: Benthic foraminiferal $^{13}C$ and accumulation rates of organic carbon: *Uvigerina peregrina* and *Cibicidoides wuellerstorfi. Paleoceanography 1*, 27–42.

Zheng, S. 1988: *The agglutinated and Porcelaneous Foraminifera of the East China Sea.* 337 pp. (In Chinese with English abstract and description of new taxa.)

Zwaan, G.J. Van der & Jorissen, F.J. 1991: Biofacial patterns in riverinduced shelf anoxia. *In* Tyson, R.V. & Pearson, T.H. (eds): *Modern and Ancient Continental Shelf Anoxia. Geological Society of London Special Publication 58*, 65–82.

# Appendix

Census data for the benthic foraminifer species in the >125 µm fraction from Hole 502A and Hole 503. Abundances are expressed as total number of benthic foraminifer fauna in each sample.

Core 502 - 1

| SITE | 502A |
| --- | --- |
| CORE | (see table) |
| SECTION | (see table) |
| INTERVAL (cm) | (see table) |
| DEPTH (mbsf) | (see table) |
| AGE (MA) | (see table) |

Species listed (rows):

- *Allomorphina pacifica*
- *Anomalinoides globulosus*
- *Astrononion gallowayi*
- *Bolivina catanensis*
- *Bolivina seminuda*
- *Bolivina subaenariensis*
- *Bulimina alazanensis*
- *Bulimina marginata*
- *Bulimina translucens*
- *Cassidulina carinata*
- *Cassidulina crassa*
- *Chilostomella oolina*
- *Chrysalogonium lanceolum*
- *Chrysalogonium longicostatum*
- *Chrysalogonium tenuicostatum*
- *Cibicidoides bradyi*
- *Cibicidoides kullenbergi*
- *Cibicidoides lobatulus*
- *Cibicidoides mundulus*
- *Cibicidoides rhodiensis*
- *Cibicidoides robertsonianus*
- *Cibicidoides wuellerstorfi*
- *Dentalina communis*
- *Dentalina cf. communis*
- *Dentalina filiformis*
- *Dentalina intorta*
- *Eggerella bradyi*
- *Entomorphinoides?* aff. *separata*
- *Epistominella exigua*
- *Eponides bradyi*
- *Eponides polius*
- *Fissurina auriculata*
- *Fissurina cf. capillosa*
- *Fissurina castrensis*
- *Fissurina clathrata*
- *Fissurina collifera*
- *Fissurina crebra*
- *Fissurina fimbriata*
- *Fissurina kerguelenensis*
- *Fissurina marginata*
- *Fissurina orbignyana*
- *Fissurina semimarginata*
- *Fissurina sulcata*
- *Fissurina wiesneri*
- *Fissurina spp.*
- *Florilus atlanticus*
- *Francesita advena*
- *Glandulina laevigata*

*Appendix* (cont.)

**Core 502 - 2**

| | | | | | | | | | | | | | | | | | | | | | | | | | | | | | | | | | | | | | | | | |
|---|---|---|---|---|---|---|---|---|---|---|---|---|---|---|---|---|---|---|---|---|---|---|---|---|---|---|---|---|---|---|---|---|---|---|---|---|---|---|---|---|
| SITE | | | | | | | | | | | | | | | | | | | | | | | | | | | | | | | | | | | | | | | | |
| CORE | 44 | 42 | 40 | 37 | | | 35 | | | 34 | | | 33 | | | | 32 | | | | 30 | | | 29 | | | | 28 | | | 27 | | | | | 26 | | | | 25 |
| SECTION | 2 | 2 | 2 | 2 | 3 | 2 | 1 | | 2 | 1 | 3 | 2 | 1 | 3 | 2 | | 1 | 3 | | 2 | 1 | 3 | 2 | 1 | 3 | 2 | | 1 | | 2 | 1 | | 3 | 2 | | 1 | 3 | | 2 | 1 |
| INTERVAL (cm) | 30-33 | 48-50 | 1-3 | 8-10 | 1-3 | 46-48 | 103-105 | 142-145 | 50-52 | 102-104 | 0-2 | 48-50 | 96-98 | 4-6 | 47-49 | 103-106 | 5-8 | 0-3 | 105-107 | 0-3 | 51-54 | 0-3 | 45-48 | 96-98 | 18-20 | 68-70 | 118-120 | 21-23 | 104-106 | 1-3 | 50-53 | 102-104 | 1-3 | 57-59 | 115-117 | 15-17 | 50-53 | 105-108 | 1-3 | 52-54 |
| DEPTH (mbsf) | 162.42 | 157.79 | 151.70 | 144.03 | 139.87 | 138.82 | 138.34 | 137.23 | 136.31 | 135.33 | 134.28 | 133.26 | 132.27 | 131.38 | 130.30 | 129.34 | 128.36 | 125.80 | 125.35 | 124.29 | 123.32 | 122.32 | 121.26 | 120.27 | 119.33 | 118.33 | 117.79 | 116.52 | 114.45 | 113.42 | 112.42 | 111.48 | 110.47 | 109.51 | 108.66 | 107.66 | 106.53 | 105.58 | 104.54 | 103.63 |
| AGE (MA) | 5.54 | 5.46 | 5.37 | 5.21 | 5.00 | 4.95 | 4.92 | 4.87 | 4.82 | 4.77 | 4.74 | 4.70 | 4.66 | 4.63 | 4.59 | 4.56 | 4.52 | 4.43 | 4.42 | 4.38 | 4.34 | 4.31 | 4.27 | 4.23 | 4.19 | 4.15 | 4.13 | 4.08 | 4.03 | 4.00 | 3.97 | 3.94 | 3.92 | 3.89 | 3.87 | 3.84 | 3.81 | 3.78 | 3.75 | 3.73 |
| *Globobulimina affinis* | – | – | – | – | – | – | – | – | – | – | – | – | – | – | – | – | – | – | – | – | – | – | – | – | – | – | – | – | – | – | – | – | – | – | – | – | – | – | – | – |
| *Globobulimina auriculata* | – | – | – | – | – | – | – | – | – | – | – | 1 | – | – | – | – | – | – | – | – | – | – | – | – | – | – | – | – | – | – | – | 1 | – | – | – | – | – | – | – | – |
| *Globobulimina pacifica* | – | – | – | – | – | – | – | – | – | – | – | – | – | – | – | 1 | – | – | – | – | – | – | – | – | – | – | – | – | – | – | – | – | – | – | – | – | – | – | – | 1 |
| *Globobulimina saubriguensis* | – | – | – | – | – | – | – | – | – | – | – | – | – | – | – | – | – | – | – | – | – | – | – | – | – | – | – | – | – | – | – | – | – | – | – | – | – | – | – | – |
| *Globocassidulina subglobosa* | – | – | 5 | 7 | 1 | – | – | – | – | – | – | – | – | – | – | – | – | 7 | – | – | – | 6 | – | – | – | 21 | – | – | – | – | 6 | – | 1 | – | – | 2 | – | – | – | – |
| *Gyroidina altiformis* | – | – | – | – | – | – | 7 | – | – | 2 | 3 | – | – | – | – | – | – | 1 | – | – | – | – | – | – | – | – | – | – | – | 2 | 2 | – | 2 | – | 2 | – | 5 | – | – | – |
| *Gyroidina neosoldanii* | 2 | 13 | 5 | 5 | 4 | 4 | 18 | 1 | 6 | 5 | 4 | 7 | 4 | 5 | – | 5 | 3 | – | 5 | 1 | 4 | 5 | 3 | 4 | 2 | 10 | 1 | 1 | 2 | 11 | 7 | 11 | 10 | 15 | 12 | 7 | 11 | 13 | 16 | 5 |
| *Gyroidinoides lamarckianus* | – | – | 5 | – | 6 | 2 | 4 | – | – | – | 2 | – | – | – | – | 1 | 1 | 2 | – | – | 2 | 6 | 1 | 2 | 1 | 5 | 1 | 1 | 2 | 4 | – | – | 2 | 3 | 2 | – | – | 1 | 4 | 6 |
| *Gyroidinoides orbicularis* | 3 | – | – | 5 | – | – | 3 | 5 | 8 | 9 | 9 | 6 | 7 | 2 | – | 7 | 7 | – | 2 | 4 | 2 | – | 3 | – | – | – | 3 | 3 | 6 | – | 7 | 7 | – | 4 | 15 | 4 | – | 17 | 12 | – |
| *Hyalinea balthica* | – | – | – | – | – | – | – | – | – | – | – | – | – | – | – | – | – | – | – | – | – | – | – | – | – | – | – | – | – | – | – | – | – | – | 2 | – | – | – | – | – |
| *Hoeglundina elegans* | – | – | – | – | – | – | – | – | 2 | – | – | – | – | – | 1 | – | – | 1 | – | – | – | 3 | – | – | – | 2 | 1 | – | – | 1 | 4 | – | – | – | – | – | – | – | – | – |
| *Karreriella bradyi* | – | – | – | – | – | – | – | – | – | – | – | – | – | – | – | – | – | – | – | – | – | – | – | – | 1 | – | – | 1 | – | – | – | – | – | – | – | – | – | – | – | – |
| *Lagena advena* | – | – | – | 1 | – | – | – | – | – | – | – | – | – | – | 1 | – | – | – | – | – | – | – | 1 | – | – | – | – | 1 | 1 | 2 | 1 | – | – | 1 | – | – | – | – | 1 | 1 |
| *Lagena alticostata* | – | – | – | – | – | – | – | – | – | – | – | – | – | – | – | – | – | – | 1 | – | – | – | – | – | – | – | – | – | – | – | – | – | – | – | – | – | – | – | – | – |
| *Lagena distoma* | – | – | – | – | – | 1 | – | – | – | – | – | – | – | – | – | – | – | – | – | – | – | – | – | – | – | – | – | – | – | – | – | – | – | – | – | – | – | – | – | – |
| *Lagena feildeniana* | – | – | – | – | – | – | – | – | – | – | – | – | – | 1 | – | – | – | – | – | – | – | – | – | – | – | – | – | – | – | – | – | – | – | – | – | – | – | – | – | – |
| *Lagena hispida* | – | – | – | – | – | – | – | – | 2 | – | 1 | – | – | 1 | – | – | – | – | – | – | 3 | – | – | – | – | – | 2 | 2 | – | – | – | – | 1 | – | – | – | – | 1 | 1 | – |
| *Lagena hispidula* | – | – | – | – | – | – | – | – | 1 | – | – | 3 | – | – | – | – | – | – | – | – | – | – | – | – | – | – | – | – | 1 | – | – | – | – | – | – | – | 2 | – | – | – |
| *Lagena meridionalis* | – | – | – | – | – | – | – | – | – | – | – | – | – | – | – | – | – | – | – | – | – | – | – | – | – | – | – | – | – | – | – | – | – | – | – | – | – | – | – | – |
| *Lagena paradoxa* | – | – | – | – | – | – | – | – | – | – | – | – | – | – | – | – | – | – | – | – | – | – | – | – | – | – | – | – | – | – | – | – | – | – | – | – | – | – | – | – |
| *Lagena striata* | – | – | – | – | – | – | – | – | – | – | – | – | – | – | – | – | – | – | – | – | – | – | 1 | – | – | – | – | – | – | – | – | – | – | – | – | – | – | – | – | – |
| *Lagena substriata* | – | – | – | – | – | – | – | – | – | – | – | – | – | – | 1 | – | – | – | – | – | – | – | – | – | – | – | – | – | – | – | – | – | – | – | – | – | – | – | – | – |
| *Lagena* sp. 1 | – | – | – | – | – | – | – | – | – | – | – | – | 1 | – | – | – | – | – | – | – | – | – | – | – | – | – | – | – | – | – | – | – | – | – | – | – | – | – | – | – |
| *Lagena* sp. 2 | – | – | – | – | – | – | – | – | – | – | – | – | – | 1 | – | – | – | – | – | – | – | – | – | – | – | – | – | – | – | – | – | – | – | – | – | – | – | – | – | – |
| *Laticarinina pauperata* | – | 4 | – | 12 | 5 | 7 | 22 | 4 | 5 | 2 | 9 | 9 | 10 | 5 | 5 | 9 | 17 | 9 | 14 | 9 | 6 | 14 | 7 | 7 | 10 | 14 | 9 | 12 | 10 | 14 | 8 | 10 | 26 | 5 | 11 | 5 | 7 | 29 | 3 | – |
| *Lenticulina atlantica* | 1 | 1 | – | 1 | – | – | – | 3 | – | – | 1 | 2 | – | – | – | 1 | 1 | 1 | 2 | – | – | 1 | – | – | 9 | – | 1 | 1 | – | – | 1 | 1 | 3 | – | 1 | – | – | 4 | – | 1 |
| *Marginulina obesa* | – | – | – | – | – | – | – | – | – | – | – | – | – | – | – | – | – | – | – | – | – | – | – | – | – | – | – | – | – | – | – | – | – | – | – | – | – | – | – | – |
| *Marginulina?* sp. | – | – | – | – | – | – | – | – | – | – | – | – | – | – | – | – | – | – | – | 1 | – | – | – | – | – | – | – | – | – | – | – | – | – | – | – | – | – | – | – | – |
| *Martinottiella communis* | – | 1 | – | – | – | – | – | 1 | – | 1 | – | – | – | – | – | – | 1 | – | 1 | 1 | – | – | – | – | 1 | – | – | – | 1 | 1 | – | – | 1 | 4 | 1 | – | – | 4 | 3 | – |
| *Martinottiella milletti* | – | – | – | – | – | – | – | – | – | – | 1 | – | – | – | – | – | – | – | – | – | – | – | – | – | – | – | – | – | – | – | – | – | – | – | – | – | – | – | – | – |
| *Melonis affinis* | 5 | 7 | – | 5 | – | 4 | – | – | 5 | 2 | 5 | 3 | – | – | – | – | 4 | – | 6 | 2 | 5 | 6 | – | – | 3 | – | 5 | – | 3 | 5 | 6 | – | – | 4 | 2 | 2 | 6 | 7 | 8 | – |
| *Melonis barleeanum* | – | – | – | 2 | 6 | 4 | 5 | – | – | – | – | 2 | – | – | – | – | – | 2 | – | – | 2 | – | 3 | 2 | – | 6 | – | – | – | 1 | – | – | 2 | 3 | – | – | 1 | 3 | 1 | 1 |
| *Melonis pompilioides* | – | – | 5 | – | – | – | 3 | – | – | – | 1 | – | – | – | – | 1 | 3 | – | 2 | 1 | – | – | 1 | – | – | 1 | – | – | 2 | 1 | 8 | 1 | – | 4 | 2 | – | – | – | – | – |
| *Monalysidium politum?* | – | – | – | – | – | – | – | – | – | – | – | – | – | – | – | – | – | – | – | – | – | – | – | – | – | – | – | – | – | – | – | – | – | – | – | – | – | – | – | – |
| *Nodosaria albatrossi* | – | – | – | – | – | – | – | – | – | – | – | – | – | – | – | – | – | – | – | – | – | – | – | – | – | – | – | – | – | – | – | – | – | – | – | – | – | – | – | 1 |
| *Nodosaria catesbyi* | – | – | – | – | – | – | – | – | – | – | – | – | – | – | – | – | – | – | – | – | – | – | – | – | – | – | – | – | – | – | – | – | – | – | – | – | – | – | – | – |
| *Nodosaria simplex* | – | – | – | – | – | – | – | – | – | – | – | – | – | – | – | – | – | – | – | – | – | – | – | – | – | – | – | – | – | – | – | – | – | – | – | – | – | – | – | – |
| *Nonion germanicum* | – | – | – | – | – | 3 | – | 4 | 1 | 1 | 3 | – | – | – | – | 2 | – | – | 1 | – | – | 4 | – | – | – | 3 | – | 1 | 5 | 9 | – | 4 | 3 | 4 | – | 2 | – | 4 | – | 2 |
| *Nummoloculina irregularis* | – | – | 1 | – | – | – | – | – | – | – | – | – | – | – | – | – | – | – | – | – | – | – | – | – | – | – | – | – | – | – | – | – | – | – | – | – | – | – | – | – |
| *Nuttallides umbonifera* | 26 | 94 | 46 | 9 | 112 | 25 | 38 | 43 | 116 | 26 | 52 | 22 | 17 | 21 | 23 | 44 | 5 | 20 | 67 | 27 | 47 | 126 | 27 | 13 | 12 | 35 | 19 | 10 | 14 | 2 | 91 | 5 | 13 | 1 | – | 3 | 3 | 3 | 2 | 48 |
| *Oolina alifera* | – | – | – | – | – | – | – | – | – | – | – | – | – | – | – | – | – | – | – | – | – | – | – | – | – | – | – | – | – | – | – | – | – | – | – | – | – | – | – | – |
| *Oolina desmophora* | – | – | – | – | – | – | – | – | – | – | – | – | – | – | – | – | – | – | – | – | – | – | – | – | – | – | – | – | – | – | – | – | – | – | – | – | – | – | – | – |
| *Oolina globosa* | – | – | – | – | – | – | – | – | – | – | – | – | – | – | – | – | 1 | – | 2 | – | – | – | – | – | – | – | – | – | – | – | – | – | – | – | – | – | – | – | – | – |
| *Oolina setosa* | – | – | – | – | – | – | – | – | – | – | – | – | – | – | – | – | – | – | – | – | – | – | – | – | – | – | – | – | – | – | – | – | – | – | – | – | – | – | – | – |
| *Oolina* sp. 1 | – | – | – | – | – | – | – | – | – | – | – | – | – | – | – | – | – | – | – | 1 | – | – | – | – | – | – | – | – | – | – | – | – | – | – | 2 | – | – | – | – | – |
| *Oridorsalis umbonatus* | 1 | 14 | 8 | 23 | 16 | 4 | 17 | 8 | 12 | 10 | 9 | 4 | 6 | 15 | 6 | 2 | 5 | 6 | 10 | 16 | 5 | 15 | 9 | 5 | 9 | 10 | 13 | 9 | 7 | 20 | 18 | 23 | 16 | 12 | 43 | 25 | 8 | 38 | 29 | 4 |
| *Orthomorphina challengeriana* | – | – | – | – | – | – | – | – | – | – | – | – | – | – | – | – | – | – | – | – | – | – | – | – | – | – | – | – | – | 2 | – | – | – | – | 1 | – | – | – | 1 | – |
| *Orthomorphina jedlitschkai* | – | – | – | – | – | – | – | – | – | – | – | – | – | – | – | – | – | – | – | – | – | – | – | – | – | – | – | – | – | – | – | – | – | – | – | – | – | – | – | – |
| *Orthomorphina* sp. 1 | – | – | 1 | – | – | – | – | – | – | – | – | – | – | – | – | – | – | – | – | – | – | – | – | – | – | – | – | – | – | – | – | – | – | – | – | – | – | – | – | – |
| *Parafissurina sublata* | – | – | – | – | – | – | – | – | – | – | – | 1 | – | – | – | – | – | – | – | – | – | – | – | – | – | – | – | – | – | – | – | – | – | – | – | – | – | – | – | – |

| | | | | | | | | | | | | | | | | | | | | | | | | | | | | | | | | | | | | | | | | |
|---|---|---|---|---|---|---|---|---|---|---|---|---|---|---|---|---|---|---|---|---|---|---|---|---|---|---|---|---|---|---|---|---|---|---|---|---|---|---|---|---|
| SITE | | | | | | | | | | | | | | | | | | | | | | | | | | | | | | | | | | | | | | | | 502A |
| CORE | | 24 | | | | 23 | | | | 22 | | | | | | 21 | | | | 20 | | | | | 19 | | | | | 17 | | | | 16 | 13 | | | | 12 | 11 |
| SECTION | 2 | 1 | CC | 3 | 2 | 1 | | 3 | 2 | 1 | CC | | 3 | | 2 | 1 | | 3 | 2 | 1 | CC | 3 | 2 | | 1 | CC | 3 | | 2 | 1 | 3 | 2 | | 1 | 1 | | 3 | 2 | 1 | CC |
| INTERVAL (cm) | 51-53 | 102-104 | 2-4 | 52-54 | 105-107 | 46-48 | 101-104 | 0-3 | 43-45 | 103-105 | 0-3 | 50-53 | 37-39 | 101-104 | 0-3 | 55-57 | 97-99 | 0-3 | 55-57 | 97-99 | 0-3 | 0-3 | 51-53 | 98-100 | 0-3 | 0-3 | 51-53 | 100-102 | 0-3 | 51-53 | 0-3 | 51-53 | 100-103 | 4-6 | 12-15 | 88-90 | 0-3 | 47-50 | 96-99 | 11-13 |
| DEPTH (mbsf) | 100.73 | 99.73 | 98.81 | 97.80 | 96.82 | 94.77 | 94.48 | 92.91 | 91.85 | 90.94 | 89.58 | 88.97 | 88.84 | 88.02 | 86.98 | 86.06 | 84.96 | 83.99 | 83.03 | 81.91 | 80.46 | 79.47 | 78.45 | 77.69 | 76.72 | 72.39 | 71.42 | 70.42 | 69.43 | 68.42 | 66.51 | 65.51 | 64.51 | 63.55 | 50.43 | 49.67 | 48.80 | 47.76 | 46.87 | 41.62 |
| AGE (MA) | 3.65 | 3.62 | 3.59 | 3.56 | 3.54 | 3.48 | 3.45 | 3.43 | 3.40 | 3.36 | 3.31 | 3.29 | 3.28 | 3.25 | 3.21 | 3.18 | 3.13 | 3.10 | 3.08 | 3.05 | 3.00 | 2.97 | 2.93 | 2.91 | 2.87 | 2.73 | 2.70 | 2.66 | 2.63 | 2.60 | 2.53 | 2.50 | 2.47 | 2.44 | 2.00 | 1.97 | 1.94 | 1.91 | 1.88 | 1.71 |
| *Globobulimina affinis* | – | – | – | – | – | – | – | – | – | – | – | – | – | – | – | – | – | – | – | – | – | – | – | – | – | 1 | – | – | – | – | – | – | – | – | – | – | – | – | – | – |
| *Globobulimina auriculata* | – | – | – | – | – | – | – | – | – | – | – | – | – | – | – | – | – | – | – | – | – | – | – | – | – | – | – | – | – | – | – | – | – | – | – | – | 11 | – | – | – |
| *Globobulimina pacifica* | – | – | – | – | 1 | – | – | – | – | – | – | – | – | – | – | – | – | – | – | – | – | – | – | – | – | – | – | – | – | – | – | – | – | – | – | – | – | – | 1 | – |
| *Globobulimina saubriguensis* | – | – | – | – | – | – | – | – | – | – | – | – | 6 | – | – | – | – | 1 | – | – | – | – | – | – | 5 | – | – | – | – | – | – | – | – | – | – | – | – | – | – | – |
| *Globocassidulina subglobosa* | – | 6 | 15 | 16 | 33 | 14 | – | 29 | 4 | 8 | 4 | – | 2 | 4 | 2 | – | – | – | – | 2 | – | – | – | – | 3 | – | – | – | – | 3 | – | – | – | 2 | – | – | 3 | – | – | – |
| *Gyroidina altiformis* | – | – | – | – | 2 | – | – | – | – | – | – | – | – | – | 1 | – | 1 | – | – | – | – | – | – | – | – | – | – | – | – | – | 2 | – | – | – | – | – | 3 | – | – | 4 |
| *Gyroidina neosoldanii* | 3 | 6 | 7 | 4 | 7 | 12 | 9 | 8 | 6 | 8 | 11 | 2 | 6 | 7 | 3 | 1 | 4 | 5 | 9 | 7 | 3 | 5 | 13 | 7 | 8 | 7 | 8 | 2 | 7 | 5 | 11 | 16 | 2 | 11 | 18 | 5 | 7 | 3 | 5 | 9 |
| *Gyroidinoides lamarckianus* | – | 2 | 4 | 3 | – | – | 3 | – | 2 | – | – | 1 | – | – | 2 | – | – | 2 | 2 | – | – | – | 2 | – | 2 | 1 | 3 | 2 | – | – | 3 | – | 2 | 3 | 5 | – | – | – | 9 | – |
| *Gyroidinoides orbicularis* | 7 | 4 | 5 | 6 | 7 | 17 | 3 | 5 | – | 9 | 9 | 2 | 9 | 4 | 6 | 5 | 9 | 10 | 8 | 6 | 9 | 24 | 14 | 7 | 12 | 20 | 14 | 13 | 22 | 5 | 4 | – | 2 | 3 | – | – | 12 | – | 9 | – |
| *Hyalinea balthica* | – | – | – | – | – | – | – | – | – | – | – | – | – | – | – | – | – | – | – | – | – | – | – | – | – | – | 1 | – | – | – | – | – | – | – | 1 | – | 20 | – | – | – |
| *Hoeglundina elegans* | – | 3 | – | – | – | – | – | – | – | – | – | 1 | – | – | – | – | – | – | – | – | – | – | – | – | – | – | – | – | – | – | – | – | – | – | – | – | 3 | – | – | – |
| *Karreriella bradyi* | – | – | – | – | – | – | – | – | – | – | – | – | – | – | – | – | – | – | – | – | – | – | – | – | – | – | – | – | – | – | – | – | – | – | – | – | – | – | – | – |
| *Lagena advena* | 1 | 1 | – | – | – | 2 | 1 | 2 | 1 | – | – | – | – | – | – | – | – | – | 2 | – | – | – | – | – | – | – | – | – | – | – | – | – | – | – | – | – | – | 1 | – | – |
| *Lagena alticostata* | – | – | – | – | – | – | – | – | 1 | – | – | – | – | – | – | – | – | – | – | – | – | – | – | – | – | – | – | – | – | – | – | – | – | – | – | – | – | – | – | – |
| *Lagena distoma* | – | – | – | – | – | – | – | – | – | – | – | – | – | – | – | – | – | – | – | – | – | – | – | – | – | – | – | – | – | – | – | – | – | – | – | – | – | – | – | – |
| *Lagena feildeniana* | – | – | – | – | – | – | – | – | – | – | – | – | – | – | – | – | – | – | – | – | – | – | – | – | – | – | – | – | – | – | – | – | – | – | – | – | – | – | – | – |
| *Lagena hispida* | – | – | – | – | – | – | – | – | – | – | – | – | – | – | – | – | – | – | – | – | – | – | – | – | 1 | – | – | – | – | – | – | – | – | – | – | – | – | – | – | – |
| *Lagena hispidula* | – | – | – | 1 | – | – | – | – | – | – | – | – | 1 | – | – | – | – | 1 | – | – | – | – | – | – | – | 1 | 1 | – | – | – | 1 | 1 | – | – | – | 1 | – | – | – | – |
| *Lagena meridionalis* | – | – | – | – | – | – | – | – | – | – | – | – | 1 | – | – | – | – | – | – | – | – | – | – | – | – | – | – | – | – | – | – | – | – | – | – | – | – | – | – | – |
| *Lagena paradoxa* | – | – | – | – | – | 2 | – | – | – | – | – | – | – | – | – | – | – | 3 | – | – | 1 | – | – | – | – | 2 | 1 | – | – | – | – | – | – | – | – | – | – | – | – | – |
| *Lagena striata* | – | – | – | – | – | – | – | – | – | – | – | 1 | – | – | – | – | – | – | – | – | – | – | – | 1 | – | – | – | 1 | – | – | – | – | – | – | – | – | – | – | 1 | – |
| *Lagena substriata* | – | – | – | – | – | – | – | – | – | – | – | – | – | – | – | – | – | – | – | – | – | – | – | – | – | – | – | – | – | – | – | – | – | – | – | – | – | – | – | – |
| *Lagena* sp. 1 | – | – | 1 | – | – | – | – | – | – | – | – | – | 1 | – | – | – | – | 1 | – | – | – | – | – | – | – | – | – | – | – | – | – | 1 | – | – | – | – | – | – | – | – |
| *Lagena* sp. 2 | – | – | – | – | – | – | – | – | – | – | – | – | – | – | – | – | – | – | – | – | – | – | – | – | – | – | – | – | – | – | – | – | – | – | – | – | – | – | – | – |
| *Laticarinina pauperata* | 17 | 3 | 4 | 11 | 8 | 2 | 5 | 1 | 4 | 6 | 13 | 9 | 3 | – | – | 2 | – | 5 | 6 | – | 8 | 6 | 15 | 3 | – | 6 | 12 | 9 | 7 | 3 | 23 | – | 1 | 2 | 4 | 1 | – | 4 | 6 | 4 |
| *Lenticulina atlantica* | 5 | 1 | – | – | – | 1 | – | – | – | 1 | 2 | 1 | 1 | – | – | – | – | 2 | 1 | – | – | – | 1 | – | 1 | 1 | 6 | – | – | 1 | – | 2 | 3 | 1 | 2 | 1 | 2 | – | 1 | 1 |
| *Marginulina obesa* | – | – | – | – | – | – | – | – | – | – | – | – | – | – | – | – | – | – | – | – | – | – | – | – | – | – | – | – | – | – | – | – | – | – | – | – | – | – | 1 | – |
| *Marginulina?* sp. | – | – | – | – | – | – | – | – | – | – | – | – | – | – | – | – | – | – | – | – | 1 | – | – | – | – | – | – | – | – | – | – | – | – | – | – | – | – | – | – | – |
| *Martinottiella communis* | 1 | – | – | – | – | – | – | – | – | 1 | – | – | – | – | – | – | – | – | – | – | 6 | – | 1 | – | – | – | 3 | – | – | – | – | – | – | – | – | – | – | – | – | – |
| *Martinottiella milletti* | – | – | – | – | – | – | – | – | – | – | – | – | – | – | – | – | – | – | – | – | – | – | – | – | – | – | – | – | – | – | – | – | – | – | – | – | – | – | – | – |
| *Melonis affinis* | 6 | 5 | 6 | 10 | 6 | 6 | 5 | 5 | 5 | 6 | 10 | 4 | – | 3 | 6 | 3 | 13 | 10 | 22 | 4 | 31 | – | 15 | – | 7 | 4 | 6 | 4 | – | 1 | 3 | 12 | 9 | 3 | 1 | – | – | 1 | – | – |
| *Melonis barleeanum* | – | – | – | – | – | – | – | – | – | – | – | 1 | – | – | – | – | – | – | 12 | – | 9 | 19 | 1 | – | – | – | 3 | – | 2 | – | – | 1 | 1 | – | – | – | 23 | 1 | – | 12 |
| *Melonis pompilioides* | – | – | – | 1 | – | 1 | 1 | 4 | 1 | – | – | – | – | 5 | 1 | – | 3 | – | 2 | 3 | 1 | 1 | 1 | 1 | 13 | 5 | 11 | – | 10 | 5 | 9 | 6 | 11 | 8 | 7 | 3 | – | 45 | 2 | 5 |
| *Monalysidium politum?* | – | – | – | – | – | – | – | – | – | – | – | – | – | – | – | – | – | – | – | – | – | – | – | – | – | – | – | – | – | – | – | – | – | – | – | – | – | – | – | – |
| *Nodosaria albatrossi* | – | – | – | – | – | – | – | – | – | – | – | – | – | – | – | – | – | – | – | – | – | 2 | – | – | – | – | – | – | – | – | – | – | – | – | – | – | – | – | – | – |
| *Nodosaria catesbyi* | – | – | – | – | – | – | – | – | – | – | – | – | – | – | – | – | – | – | – | – | – | – | – | – | – | – | – | – | – | – | – | – | – | – | – | – | – | – | – | – |
| *Nodosaria simplex* | – | – | – | – | – | – | – | – | – | – | – | – | – | – | – | – | – | – | – | – | – | – | – | 1 | – | – | – | – | – | – | – | – | – | – | – | – | – | – | – | – |
| *Nonion germanicum* | 1 | 2 | – | 2 | 7 | 4 | 3 | 5 | 3 | 3 | – | 3 | 16 | 8 | – | 1 | 2 | 6 | 8 | 4 | 4 | – | 7 | 2 | 13 | 3 | 4 | 3 | 3 | 4 | – | 1 | 1 | 4 | 9 | – | – | – | – | – |
| *Nummoloculina irregularis* | – | – | – | – | – | – | – | – | – | – | – | – | – | – | – | – | – | – | – | – | – | – | – | – | – | – | – | – | – | – | – | – | – | – | – | – | – | – | – | – |
| *Nuttallides umbonifera* | 89 | 139 | 177 | 90 | 140 | 185 | 136 | 159 | 185 | 76 | 170 | 2 | 123 | 167 | 189 | 106 | 202 | 166 | 140 | 266 | 104 | 189 | 131 | 63 | – | 158 | 53 | 74 | 41 | 60 | 119 | 84 | 149 | 131 | 62 | 141 | 132 | 41 | 241 | 174 |
| *Oolina alifera* | – | – | – | – | – | – | – | – | – | – | – | – | – | – | – | – | – | – | – | – | – | – | – | – | – | – | – | – | – | – | 1 | 1 | – | – | – | 1 | – | – | – | – |
| *Oolina desmophora* | – | – | – | – | – | – | – | – | – | – | – | – | – | – | – | – | – | – | – | – | – | – | – | – | – | – | – | – | – | – | – | – | – | – | – | – | – | – | – | – |
| *Oolina globosa* | – | 1 | – | – | – | – | – | – | – | – | – | – | – | – | – | – | – | – | – | – | – | – | – | – | – | – | – | – | – | – | – | – | – | – | – | – | – | – | – | – |
| *Oolina setosa* | – | – | – | – | – | – | – | – | – | – | – | – | – | – | – | – | – | – | – | – | – | – | – | – | – | 1 | – | – | – | – | 1 | 1 | – | – | – | – | – | – | – | – |
| *Oolina* sp. 1 | – | – | – | – | – | – | – | – | – | – | 1 | – | – | – | – | – | – | – | – | – | – | – | – | – | – | – | – | – | – | – | – | – | – | – | – | – | – | – | – | – |
| *Oridorsalis umbonatus* | 43 | 15 | 22 | 36 | 16 | 2 | 18 | 7 | 27 | 26 | 30 | 10 | – | 15 | 4 | 11 | – | 7 | 20 | 21 | 11 | 4 | 14 | 27 | 16 | 43 | 24 | 6 | 31 | 20 | 47 | 33 | 18 | 25 | 26 | 21 | 4 | 22 | 9 | – |
| *Orthomorphina challengeriana* | – | – | – | – | 1 | – | – | 1 | – | – | – | – | – | 2 | 2 | 2 | – | – | – | – | – | – | – | – | – | – | – | – | – | – | – | – | – | – | – | – | – | – | – | – |
| *Orthomorphina jedlitschkai* | – | – | – | – | – | – | – | – | – | – | – | – | – | – | – | – | – | – | – | – | – | – | – | – | – | 2 | – | – | – | – | – | – | – | – | – | – | – | – | – | – |
| *Orthomorphina* sp. 1 | – | – | – | – | – | – | – | – | – | – | – | – | – | – | – | – | – | – | – | – | – | – | – | – | – | – | – | – | – | – | 1 | – | – | – | – | – | – | – | – | – |
| *Parafissurina sublata* | – | – | – | – | – | – | – | – | – | – | – | – | – | – | – | – | – | – | – | – | – | – | – | – | – | – | – | – | – | – | – | – | – | – | – | – | – | – | – | – |

*Appendix* (cont.)

Core 502 - 3

| | SITE |
|---|---|
| 502A | CORE |
| | SECTION |
| | INTERVAL (cm) |
| | DEPTH (mbsf) |
| | AGE (MA) |

*Parafissurina tectulostoma*
*Parafissurina tricarinata*
*Parafissurina uncifera*
*Parafissurina* sp. 1
*Pleurostomella acuminata*
*Pleurostomella alternans*
*Pleurostomella brevis*
*Pleurostomella recens*
*Pleurostomella subnodosa*
*Polymorphinidae formae fistulosae*
*Pseudonodosaria* sp. 1
*Pullenia bulloides*
*Pullenia quinqueloba*
*Pullenia riveroi?*
*Pullenia subcarinata*
*Pyrgo depressa*
*Pyrgo murrhina*
*Pyrgo oblonga*
*Pyrulina extensa*
*Pyrulina fusiformis*
*Pyrulina gutta*
*Pyrulina* sp. 1
*Pyrulina* sp. 2
*Pyrulinoides* sp. 1
*Pyrulinoides* sp. 2
*Pyrulinoides* sp. 3
*Quinqueloculina weaveri*
*Quinqueloculina venusta*
*Recurvoides scitulus*
*Rutherfordoides bradyi*
*Rutherfordoides tenuis*
*Saracenaria italica*
*Saracenaria latifrons*
*Sigmoilina tenuis*
*Sigmoilopsis schlumbergeri*
*Siphonodosaria abyssorum*
*Siphonodosaria consobrina*
*Siphonodosaria lepidula*
*Siphonodosaria simplex*
*Siphotextularia catenata*
*Stilostomella* sp.
*Triloculina tricarinata*
*Uvigerina auberiana*
*Uvigerina canariensis*
*Uvigerina hispida*
*Uvigerina hollicki*
*Uvigerina mantaensis*
*Uvigerina peregrina*
*Valvulineria humilis*
UNIDENTIFIED

*Appendix* (cont.)

**Core 503 -1**

| SITE | CORE | SECTION | INTERVAL (cm) | DEPTH (mbsf) | AGE (MA) | *Allomorphina pacifica* | *Anomalinoides globulosus* | *Astrononion gallowayi* | *Bolivina catanensis* | *Bolivina seminuda* | *Bolivina subaenariensis* | *Bulimina alazanensis* | *Bulimina marginata* | *Bulimina translucens* | *Cassidulina carinata* | *Cassidulina crassa* | *Chilostomella oolina* | *Chrysalogonium lanceolum* | *Chrysalogonium longicostatum* | *Chrysalogonium tenuicostatum* | *Cibicidoides bradyi* | *Cibicidoides kullenbergi* | *Cibicidoides lobatulus* | *Cibicidoides mundulus* | *Cibicidoides rhodiensis* | *Cibicidoides robertsonianus* | *Cibicidoides wuellerstorfi* | *Dentalina communis* | *Dentalina* cf. *communis* | *Dentalina filiformis* | *Dentalina intorta* | *Eggerella bradyi* | *Entomorphinoides?* aff. *separata* | *Epistominella exigua* | *Eponides bradyi* | *Eponides polius* | *Fissurina auriculata* | *Fissurina* cf. *capillosa* | *Fissurina castrensis* | *Fissurina clathrata* | *Fissurina collifera* | *Fissurina crebra* | *Fissurina fimbriata* | *Fissurina kerguelenensis* | *Fissurina marginata* | *Fissurina orbignyana* | *Fissurina semimarginata* | *Fissurina sulcata* | *Fissurina wiesneri* | *Fissurina spp.* | *Florilus atlanticus* | *Francesita advena* | *Glandulina laevigata* | *Globobulimina affinis* | *Globobulimina auriculata* | *Globobulimina pacifica* | *Globobulimina saubriguensis* |
|---|---|---|---|---|---|---|---|---|---|---|---|---|---|---|---|---|---|---|---|---|---|---|---|---|---|---|---|---|---|---|---|---|---|---|---|---|---|---|---|---|---|---|---|---|---|---|---|---|---|---|---|---|---|---|---|---|---|
| 503B | 8 | 1 | 61-63 | 29.82 | 1.66 | – | – | – | – | – | – | – | – | – | – | – | – | – | – | – | 1 | 1 | – | – | – | – | 2 | – | – | – | – | – | – | 4 | – | – | – | – | – | – | – | 1 | – | – | 1 | – | – | – | – | – | – | – | – | – | – | – | – |
| | 8 | 3 | 116-118 | 33.28 | 1.81 | – | – | – | – | – | – | – | – | – | – | – | – | 1 | – | – | 2 | 3 | – | 3 | – | – | 3 | – | – | – | – | – | – | 5 | – | – | – | – | – | – | – | – | – | – | – | – | – | – | – | 1 | – | – | – | – | – | – | – |
| | 10 | 1 | 12-14 | 38.13 | 2.01 | – | – | 1 | – | – | – | – | – | – | – | – | – | – | – | – | – | 5 | – | – | – | – | 8 | – | – | – | – | – | – | – | – | – | – | 1 | – | – | – | – | – | – | 3 | – | – | – | – | – | – | – | – | – | – | – | – |
| | | | 112-114 | 39.13 | 2.05 | – | – | – | – | – | – | – | – | – | – | – | – | – | – | – | 1 | – | – | – | – | – | 1 | – | – | – | – | 3 | – | 6 | – | – | – | – | – | – | – | – | – | – | 1 | – | – | – | 1 | – | – | – | – | – | – | – | – |
| | | 2 | 10-12 | 39.58 | 2.07 | – | – | – | – | – | – | – | – | – | – | – | – | – | – | – | 2 | – | – | – | – | – | 4 | – | – | – | – | – | – | 9 | – | 6 | – | – | – | – | – | – | – | – | – | – | – | – | – | 1 | – | – | – | – | – | – | – |
| | | | 60-62 | 40.08 | 2.09 | – | – | – | – | – | – | – | – | – | – | – | 1 | – | – | – | – | 2 | – | 1 | – | 1 | 4 | – | – | – | – | – | – | 7 | – | – | – | 1 | – | – | – | – | – | – | – | – | – | – | – | 1 | – | – | – | – | – | – | – |
| | | 3 | 10-13 | 41.08 | 2.14 | – | – | – | – | – | – | – | – | – | – | – | 1 | – | – | – | – | 3 | – | – | – | – | – | – | – | – | – | 1 | – | 3 | – | – | 3 | – | – | – | – | – | – | – | – | – | – | – | – | 2 | – | – | – | – | – | – | – |
| | 11 | 2 | 11-13 | 43.79 | 2.25 | – | – | – | – | – | – | 1 | – | – | – | – | – | – | – | – | – | 13 | – | – | – | – | 7 | – | – | – | – | – | – | – | – | – | – | – | – | – | – | – | – | – | 1 | – | – | – | – | – | – | 1 | – | – | – | – | – |
| | | | 61-63 | 44.29 | 2.27 | – | – | – | – | – | – | – | – | – | – | – | 1 | – | – | – | 3 | 4 | – | – | – | – | – | – | – | – | – | 3 | – | 9 | – | 12 | – | – | – | – | – | – | – | – | – | – | – | – | – | – | – | – | – | – | – | – | – |
| | | | 112-114 | 44.80 | 2.29 | – | – | – | – | – | – | – | – | – | – | – | 1 | – | – | – | 3 | 1 | – | 1 | – | 16 | 21 | – | – | – | – | 3 | – | 6 | – | – | – | 1 | – | – | – | – | – | – | – | – | – | – | – | – | – | – | – | – | – | – | – |
| | | 3 | 62-64 | 45.80 | 2.33 | – | – | – | – | – | – | – | – | – | – | – | 2 | 1 | – | – | 3 | – | – | – | – | 1 | 6 | – | – | – | – | – | – | 2 | – | – | – | – | – | – | – | – | – | – | – | – | – | – | – | – | – | – | – | – | – | – | – |
| | 12 | 1 | 7-9 | 46.88 | 2.38 | – | – | – | – | – | 1 | – | – | – | – | – | – | – | – | – | – | 3 | – | – | – | – | 6 | – | – | – | – | 1 | – | 3 | – | – | 1 | 1 | – | – | – | – | – | – | 2 | – | – | – | – | – | – | 1 | – | – | – | – | – |
| | | | 106-108 | 47.87 | 2.42 | – | – | – | – | – | – | – | – | – | – | – | 3 | – | – | – | – | 4 | – | – | – | – | 5 | – | – | – | – | – | – | 3 | – | – | 1 | – | – | – | – | – | – | – | – | – | – | – | – | – | – | – | – | – | – | – | 2 |
| | | 2 | 57-59 | 48.88 | 2.46 | – | – | – | – | – | – | – | – | 1 | – | – | 2 | – | – | – | – | 7 | – | 1 | – | – | 11 | – | – | – | – | 1 | – | 19 | 1 | – | – | – | – | – | – | – | – | – | 1 | – | – | – | – | – | – | – | – | – | – | – | – |
| | 13 | 1 | 22-24 | 51.43 | 2.57 | – | – | – | – | – | – | 1 | – | – | – | – | – | – | – | – | – | 2 | – | – | – | – | 11 | – | – | – | – | 1 | – | 7 | – | 4 | – | – | – | – | – | – | – | – | – | – | – | – | – | – | – | – | – | – | – | – | – |
| | | | 75-77 | 51.96 | 2.59 | – | – | – | – | – | – | 2 | – | – | – | – | – | – | – | – | – | – | – | – | – | – | 8 | – | – | – | – | – | – | 23 | – | – | – | – | – | – | – | – | – | – | – | – | 1 | – | – | 1 | – | – | – | – | – | – | – |
| | | | 126-128 | 52.47 | 2.61 | – | – | – | – | – | – | 1 | – | – | – | – | 1 | – | – | – | – | 4 | – | – | – | 4 | 9 | – | – | – | – | – | – | 10 | 2 | – | – | – | – | – | – | – | – | – | – | – | 1 | – | – | – | – | – | – | – | – | – | – |
| | | 2 | 81-84 | 53.47 | 2.63 | – | – | – | – | – | – | – | – | – | – | – | – | – | – | – | – | 4 | – | – | – | – | 6 | – | – | – | – | – | – | 4 | 1 | – | – | – | – | – | – | – | – | – | – | – | – | – | – | – | – | – | – | – | – | – | – |
| | | 3 | 37-39 | 54.53 | 2.69 | – | – | – | – | – | – | – | – | – | – | – | – | – | – | – | – | – | – | – | – | – | 11 | – | – | – | – | – | – | 12 | 4 | – | 1 | – | – | – | – | – | – | – | 1 | – | – | – | – | – | – | – | – | – | – | – | 1 |
| | 14 | 1 | 78-80 | 56.39 | 2.78 | – | – | – | – | – | – | 1 | – | – | – | – | – | – | – | – | – | 4 | – | – | – | – | 6 | – | – | – | – | 1 | – | 3 | – | 4 | – | – | – | – | – | – | – | – | – | – | – | – | – | – | – | – | – | – | – | – | – |
| | | | 128-130 | 56.89 | 2.79 | – | 1 | – | – | – | – | – | – | – | – | – | – | 1 | – | – | – | 2 | – | – | – | 2 | 9 | – | – | – | – | – | – | 4 | – | – | – | – | – | – | – | – | – | – | – | – | 1 | – | – | – | – | – | – | – | 2 | – | – |
| | | 2 | 128-130 | 58.00 | 2.84 | – | 1 | – | – | – | – | – | – | – | – | – | – | – | – | – | – | 1 | – | – | – | 1 | 7 | 1 | – | – | – | – | – | 4 | – | – | – | – | – | – | – | – | – | – | – | – | – | – | – | – | – | – | – | – | – | – | – |
| | | 3 | 28-30 | 58.51 | 2.87 | – | – | – | – | – | – | – | – | – | – | – | 6 | – | – | – | – | 1 | – | – | – | – | 3 | – | – | – | – | – | – | – | – | – | – | – | 1 | – | – | – | – | – | 1 | 1 | – | – | – | – | – | – | – | – | – | – | – |
| | | | 77-79 | 59.00 | 2.89 | – | – | – | – | 1 | – | – | – | – | – | – | – | – | – | – | 1 | 1 | – | 3 | – | 3 | 6 | – | – | – | – | – | – | 2 | – | – | – | – | – | – | – | – | – | – | – | – | – | – | – | – | – | – | – | – | – | – | – |
| | 15 | 2 | 61-63 | 61.77 | 2.99 | – | – | – | – | – | – | – | – | – | – | – | 4 | – | – | – | 1 | 5 | – | – | – | – | 8 | – | – | – | – | – | – | 7 | – | – | – | 1 | – | – | – | – | – | – | – | – | – | – | – | – | – | – | – | – | – | – | – |
| | | 3 | 11-13 | 62.77 | 3.03 | – | – | – | – | 1 | – | 1 | – | – | – | – | 3 | – | – | – | 1 | 3 | – | – | – | – | 7 | – | – | – | – | – | – | 3 | – | – | – | 1 | – | – | – | – | – | – | – | 1 | – | – | – | – | – | – | – | – | – | – | – |
| | | | 61-63 | 63.27 | 3.07 | – | – | – | – | – | – | – | – | – | – | – | 1 | – | – | – | – | 2 | – | – | – | 2 | 3 | – | – | – | – | – | – | 10 | 2 | – | – | – | – | – | – | – | 1 | – | – | – | – | – | – | – | – | – | – | – | – | – | – |
| | 16 | 1 | 44-46 | 64.85 | 3.11 | – | – | – | – | – | – | – | – | – | – | – | – | – | – | – | – | 3 | – | – | – | 2 | 2 | – | – | – | – | 2 | – | 9 | – | – | – | – | – | – | – | 1 | – | – | – | – | 1 | – | – | – | – | – | – | – | – | – | – |
| | | | 96-98 | 65.37 | 3.13 | – | – | – | – | – | – | – | – | – | – | – | – | – | – | – | – | 3 | – | – | – | 5 | 20 | – | – | – | – | – | – | 18 | – | – | – | – | – | – | – | – | – | – | – | – | 1 | – | – | – | – | – | – | – | – | – | – |
| | | | 146-148 | 65.87 | 3.14 | – | – | – | – | – | – | – | – | – | – | – | – | – | – | – | – | 3 | – | – | – | 4 | 9 | – | – | – | – | – | – | 6 | 2 | – | – | – | – | – | – | – | – | – | – | – | – | – | – | – | – | – | – | – | – | – | – |
| | | 2 | 100-102 | 66.75 | 3.17 | – | – | – | – | – | – | – | – | 1 | – | – | – | 1 | – | – | – | 4 | – | 1 | – | – | – | – | – | – | – | – | – | 4 | – | – | – | – | – | – | – | – | – | – | – | – | – | – | – | – | – | – | – | – | – | – | – |
| | | 3 | 1-3 | 67.26 | 3.19 | – | – | – | – | – | – | – | – | – | – | – | – | – | – | – | – | 8 | – | 2 | – | 6 | 9 | – | – | – | – | – | – | – | – | – | – | – | – | – | – | – | – | – | – | – | – | – | – | – | – | – | – | – | – | – | – |
| | | | 54-56 | 67.79 | 3.20 | – | – | – | – | – | – | – | – | – | – | – | – | 1 | – | – | – | – | – | – | – | – | 5 | – | – | – | – | – | – | 3 | 3 | – | – | – | – | – | – | – | – | – | – | – | – | – | – | – | – | – | – | – | – | – | – |
| | | | 102-104 | 68.27 | 3.22 | – | – | – | – | – | – | – | – | – | – | – | – | – | – | – | – | 4 | – | 2 | – | 2 | 5 | – | – | – | – | – | – | – | 3 | – | – | – | – | – | – | – | – | – | – | – | – | – | – | – | – | – | – | – | – | – | – |
| | 17 | 1 | 14-16 | 68.95 | 3.28 | – | – | – | – | – | – | – | – | – | – | – | 1 | – | – | – | – | – | – | – | – | 1 | 3 | – | – | – | – | – | – | 25 | 2 | – | – | 1 | 1 | – | – | – | – | – | 3 | 2 | – | – | – | – | – | – | – | – | – | – | – |
| | | | 62-64 | 69.43 | 3.29 | – | – | – | – | – | – | – | – | – | – | – | 9 | – | – | – | – | – | – | – | – | 3 | – | – | – | – | – | – | – | 11 | – | 3 | – | – | – | – | – | – | – | – | 3 | – | – | – | – | 1 | – | – | – | – | – | – | – |
| | | | 118-120 | 69.99 | 3.31 | – | – | – | – | – | – | – | – | – | – | – | 4 | – | – | – | – | – | – | – | – | – | 3 | – | – | – | – | – | – | 5 | – | – | – | – | – | – | – | – | – | – | 2 | – | – | – | – | – | – | – | – | – | – | – | – |
| | | 2 | 22-24 | 70.44 | 3.33 | – | – | – | – | – | – | – | – | – | – | – | 6 | – | – | – | – | – | – | – | – | – | 3 | – | – | – | – | – | – | – | – | – | – | 1 | – | – | – | – | – | – | – | – | – | – | – | – | – | – | – | – | – | – | – |
| | | | 74-76 | 70.96 | 3.34 | – | – | – | – | – | – | – | – | – | – | – | – | – | – | – | – | 3 | – | – | – | 2 | 3 | – | – | – | – | – | – | 9 | – | – | – | – | – | – | – | – | – | – | 1 | – | – | – | – | – | – | – | – | – | 6 | – | – |
| | | | 126-128 | 71.48 | 3.36 | – | 3 | – | – | – | – | – | – | – | – | – | 8 | 1 | – | – | – | 3 | – | – | – | – | 8 | – | – | – | – | – | – | 12 | – | 4 | – | – | – | – | – | – | – | – | 2 | – | – | – | – | 1 | – | – | – | – | – | – | – |
| | | 3 | 31-33 | 72.03 | 3.38 | 1 | – | – | – | – | – | – | – | – | – | – | – | – | – | – | – | – | – | – | – | – | 3 | – | – | – | – | – | – | 5 | 2 | – | – | – | – | – | – | – | – | – | – | – | – | – | – | – | – | 2 | 1 | – | – | – | – |
| | | | 85-87 | 72.61 | 3.40 | – | – | – | – | – | – | – | – | – | – | – | 1 | – | – | – | – | – | – | – | – | – | 4 | – | – | – | – | 1 | – | 1 | 3 | – | – | – | – | – | – | – | – | – | – | – | – | – | – | – | – | – | – | – | – | – | – |
| | 18 | 1 | 13-15 | 73.34 | 3.42 | – | – | – | – | – | – | – | – | – | – | – | 3 | – | – | – | 1 | – | – | – | – | 6 | – | – | – | – | – | 1 | – | 13 | 4 | – | – | – | – | – | – | – | – | – | 2 | – | – | – | – | – | – | – | – | – | – | – | – |
| | | | 61-63 | 73.82 | 3.44 | – | – | – | – | – | – | – | – | – | – | – | 2 | – | – | – | – | – | – | – | – | – | 3 | – | – | – | – | – | – | – | – | 3 | – | – | – | – | – | – | – | – | – | – | – | – | – | – | – | – | – | – | – | – | – |
| | | | 111-113 | 74.32 | 3.45 | – | 1 | – | – | – | – | – | – | – | – | – | 5 | 2 | – | – | – | 4 | – | – | – | – | – | – | – | – | – | 1 | – | 16 | – | – | – | – | – | – | – | – | – | – | 3 | – | – | – | – | – | – | – | – | – | – | – | – |
| | | 2 | 11-13 | 74.82 | 3.46 | – | – | – | – | – | – | – | – | – | – | – | – | – | – | 1 | 1 | – | – | – | – | 1 | 9 | – | – | – | – | – | – | 18 | – | 3 | – | – | – | – | – | – | – | – | 1 | – | – | – | – | 1 | – | – | – | – | – | – | – |
| | | | 64-66 | 75.35 | 3.46 | – | – | – | – | – | – | – | – | – | – | – | 3 | – | – | – | – | – | – | – | – | – | – | – | – | – | – | – | – | 3 | 1 | – | – | – | – | – | – | – | – | – | 3 | – | – | – | – | – | – | – | – | – | – | – | – |
| | | 3 | 14-16 | 76.36 | 3.47 | – | – | – | – | – | – | – | – | – | – | – | – | – | – | – | – | 1 | – | – | – | – | 1 | 1 | – | – | – | – | – | 6 | – | – | – | – | – | – | – | – | – | – | 2 | – | – | – | – | – | – | – | – | – | – | – | – |
| | | | 113-115 | 77.35 | 3.50 | – | – | – | – | – | – | – | – | – | – | – | – | – | – | – | – | 2 | – | – | – | 1 | – | – | – | – | – | – | – | 6 | – | – | – | 1 | 1 | – | – | – | – | – | 1 | – | – | – | – | – | – | – | – | – | – | – | – |
| | 20 | 1 | 60-62 | 81.77 | 3.63 | – | – | – | – | – | – | – | – | – | – | – | – | – | – | – | – | – | – | – | – | – | – | – | – | – | – | – | – | – | – | – | – | – | – | – | – | – | – | – | – | – | – | – | – | – | – | – | – | – | – | – | – |
| | 21 | 1 | 56-58 | 86.97 | 3.79 | – | – | – | – | – | – | – | – | – | – | – | – | – | – | – | – | – | – | – | – | – | – | 1 | – | – | – | – | – | – | – | – | – | – | 1 | – | – | – | – | – | 2 | – | – | – | – | – | – | – | – | – | – | – | – |
| | | 2 | 55-57 | 88.41 | 3.84 | – | – | – | – | – | – | – | – | – | – | – | – | – | – | – | – | 3 | – | – | – | 2 | – | 1 | – | – | – | – | – | 3 | – | – | – | – | – | – | – | – | – | – | 2 | – | – | – | – | – | – | – | – | – | – | – | – |
| | | 3 | 12-14 | 89.48 | 3.87 | – | – | – | – | – | – | – | – | – | – | – | – | – | – | – | – | 2 | – | – | – | – | – | – | – | – | – | – | – | 18 | – | – | – | – | – | – | – | – | – | – | – | – | – | – | – | – | – | – | – | – | – | – | – |
| | 22 | 1 | 56-58 | 91.37 | 3.92 | – | – | – | – | – | – | – | – | – | – | – | – | – | – | 1 | 1 | – | – | – | – | – | 3 | – | – | – | – | – | – | 3 | – | 1 | – | – | – | – | – | – | – | – | 1 | – | – | – | – | – | – | – | – | – | – | – | – |
| | | 2 | 111-113 | 93.17 | 3.95 | – | – | – | – | – | – | – | – | – | – | – | 4 | – | – | – | 2 | 2 | – | – | – | – | – | – | – | – | – | – | – | 7 | – | – | – | – | – | – | – | – | – | – | 1 | – | – | – | – | 3 | – | – | – | – | – | – | – |
| | 24 | 2 | 11-13 | 100.77 | 4.11 | – | – | – | – | – | – | – | – | – | – | – | 7 | 1 | – | – | 8 | – | – | – | – | – | 6 | – | – | – | – | – | – | 12 | 14 | – | – | – | – | – | – | – | – | – | – | – | – | – | – | 2 | – | – | – | – | – | – | – |
| | 25 | 2 | 133-135 | 106.38 | 4.21 | – | – | – | – | – | – | – | – | – | – | – | 6 | – | – | – | – | 2 | – | 2 | – | 2 | – | – | – | – | – | – | – | 2 | – | – | – | – | – | – | – | – | – | – | – | – | – | – | – | – | – | 2.0 | – | – | – | – | – |
| | 26 | 1 | 8-10 | 108.49 | 4.25 | – | – | – | – | – | – | – | – | – | – | – | 5 | – | – | – | – | 1 | – | 1 | – | – | 3 | 1 | – | – | – | – | – | 3 | – | – | – | – | – | – | – | – | – | – | – | – | – | – | – | – | – | – | – | – | – | – | – |
| | | | 108-110 | 109.49 | 4.27 | – | 1 | – | – | – | – | – | – | – | – | – | 6 | – | – | – | – | 2 | – | 3 | – | – | – | 1 | – | – | – | 1 | – | 3 | – | – | – | – | – | – | – | – | – | – | – | – | – | – | – | 2 | – | – | – | – | – | – | – |
| | | 2 | 67-69 | 109.93 | 4.28 | – | – | – | – | – | – | – | – | – | – | – | 11 | – | – | – | – | 1 | – | 6 | – | – | – | 1 | – | – | – | 2 | – | 2 | 1 | – | – | – | – | – | – | – | – | – | – | – | – | – | – | 3 | – | 1 | – | 2 | – | – | – |
| | | | 120-122 | 110.46 | 4.30 | – | – | – | – | – | – | – | – | – | – | – | – | 1 | – | – | 2 | – | – | – | – | – | 4 | – | – | – | – | 4 | – | – | – | – | – | – | – | – | – | – | – | 1 | 4 | 1 | 2 | – | – | – | – | – | – | – | – | – | – |
| 503A | 29 | 2 | 61-63 | 112.72 | 4.80 | – | – | – | – | – | – | – | – | – | – | – | – | – | – | – | – | – | – | – | – | 2 | 3 | – | – | – | – | – | – | – | – | 4 | – | – | – | – | – | – | – | – | – | – | – | – | – | – | – | – | – | – | – | – | – |
| | 30 | 2 | 41-43 | 126.75 | 4.90 | – | – | – | – | – | – | – | – | – | – | – | 3 | – | – | – | 2 | 8 | – | 1 | – | – | 14 | – | 1 | – | – | – | – | 39 | 23 | – | – | – | – | – | – | – | – | 1 | – | 1 | – | – | – | – | – | – | – | – | – | – | – |
| | 31 | 3 | 71-73 | 133.07 | 5.07 | – | – | – | – | – | – | – | – | – | – | – | 1 | – | – | – | – | 4 | – | 1 | – | 1 | 3 | – | – | – | – | – | – | 6 | – | – | 1 | – | – | – | – | – | – | – | – | 1 | – | – | – | – | – | – | – | – | – | – | – |

Appendix (cont.)

Core 503 - 2

| | | SITE |
|---|---|---|
| 503A | 503B | |

**SITE:** 503A ... 503B

**CORE:** 31 30 29 | 26 25 24 | 22 | 21 20 | 18 | 17 | 16 15 | 14 | 13 | 12 | 11 | 10 8

**SECTION:** 3 2 2 | 2 1 2 2 1 2 1 | 3 | 2 1 3 | 2 | 1 | 3 2 | 1 | 3 2 | 3 | 2 1 2 3 | 2 | 2 1 3 1

**INTERVAL (cm):** 61-63 41-43 71-73 | 120-122 67-69 8-10 108-110 | 11-13 133-135 | 56-58 55-57 113-115 14-16 64-66 11-13 | 61-63 13-15 85-87 31-33 126-128 74-76 22-24 118-120 14-16 62-64 102-104 54-56 1-3 100-102 146-148 96-98 44-46 11-13 77-79 28-30 128-130 126-128 78-80 37-39 81-84 75-77 22-24 57-59 7-9 106-108 62-64 112-114 11-13 61-63 10-12 60-62 112-114 11-13 16-118 61-63

**DEPTH (mbsf):** 112.72 126.77 133.07 | 110.46 109.93 109.49 108.38 106.77 | 100.77 91.37 89.48 | 88.41 86.97 81.77 76.36 76.35 73.84 | 73.82 72.61 71.48 70.96 70.44 69.99 69.43 68.95 68.27 67.26 66.75 65.87 64.85 62.77 62.11 61.77 59.59 59.00 58.51 58.00 56.39 54.57 53.47 52.47 51.96 51.43 48.88 47.87 46.88 45.80 44.80 43.79 41.08 40.08 39.58 39.13 38.13 33.28 29.82 20.81

**AGE (MA):** 4.80 4.90 5.07 | 4.28 4.25 4.21 4.17 4.11 | 3.95 3.84 3.79 | 3.65 3.57 3.47 3.46 3.44 3.40 | 3.38 3.36 3.34 3.33 3.31 3.29 3.28 3.25 3.22 3.20 3.19 3.17 3.13 3.07 3.05 3.03 2.89 2.87 2.84 2.80 2.78 2.63 2.61 2.59 2.57 2.49 2.46 2.38 2.33 2.27 2.20 2.14 2.09 2.07 2.05 2.01 1.81 1.66

| Species |
|---|
| *Globocassidulina subglobosa* |
| *Gyroidina altiformis* |
| *Gyroidina neosoldanii* |
| *Gyroidinoides lamarckianus* |
| *Gyroidinoides orbicularis* |
| *Hyalinea balthica* |
| *Hoeglundina elegans* |
| *Karreriella bradyi* |
| *Lagena advena* |
| *Lagena alticostata* |
| *Lagena distoma* |
| *Lagena feildeniana* |
| *Lagena hispida* |
| *Lagena hispidula* |
| *Lagena meridionalis* |
| *Lagena paradoxa* |
| *Lagena striata* |
| *Lagena substriata* |
| *Lagena* sp. 1 |
| *Lagena* sp. 2 |
| *Laticarinina pauperata* |
| *Lenticulina atlantica* |
| *Marginulina obesa* |
| *Marginulina?* sp. |
| *Martinottiella communis* |
| *Martinottiella milletti* |
| *Melonis affinis* |
| *Melonis barleeanum* |
| *Melonis pompilioides* |
| *Monalysidium politum?* |
| *Nodosaria albatrossi* |
| *Nodosaria catesbyi* |
| *Nodosaria simplex* |
| *Nonion germanicum* |
| *Nummoloculina irregularis* |
| *Nuttallides umbonifera* |
| *Oolina alifera* |
| *Oolina desmophora* |
| *Oolina globosa* |
| *Oolina setosa* |
| *Oolina* sp. 1 |
| *Oridorsalis umbonatus* |
| *Orthomorphina challengeriana* |
| *Orthomorphina jedlitschkai* |
| *Orthomorphina* sp. 1 |
| *Parafissurina sublata* |
| *Parafissurina tectulostoma* |
| *Parafissurina tricarinata* |
| *Parafissurina uncifera* |
| *Parafissurina* sp. 1 |
| *Pleurostomella acuminata* |
| *Pleurostomella alternans* |
| *Pleurostomella brevis* |

*Appendix* (cont.)

**Core 503 - 3**

| SITE | 503A | | | | | | | | | | | | | | | |
|---|---|---|---|---|---|---|---|---|---|---|---|---|---|---|---|---|
| CORE | 31 | 30 | 29 | | | | 26 | 25 | 24 | | 22 | | | 21 | 20 | |
| SECTION | 3 | 2 | 2 | | 2 | | 1 | 2 | 2 | 2 | 1 | 3 | 2 | 1 | 1 | |
| INTERVAL (cm) | 71-73 | 41-43 | 61-63 | 120-122 | 67-69 | 108-110 | 8-10 | 133-135 | 11-13 | 111-113 | 56-58 | 12-14 | 55-57 | 56-58 | 60-62 | 113-115 |
| DEPTH (mbsf) | 133.07 | 126.75 | 112.72 | 110.46 | 109.93 | 109.49 | 108.49 | 106.38 | 100.77 | 93.17 | 91.37 | 89.48 | 88.41 | 86.97 | 81.77 | 77.35 |
| AGE (MA) | 5.07 | 4.90 | 4.80 | 4.30 | 4.28 | 4.27 | 4.25 | 4.21 | 4.11 | 3.95 | 3.92 | 3.87 | 3.84 | 3.79 | 3.63 | 3.50 |
| *Pleurostomella recens* | – | – | – | – | – | 1 | – | – | 1 | – | – | 4 | 2 | – | – | – |
| *Pleurostomella subnodosa* | 1 | – | – | – | – | – | – | – | – | – | 1 | – | – | – | – | – |
| *Polymorphinidae formae fistulosae* | – | – | – | – | – | – | – | – | – | – | – | – | – | – | – | – |
| *Pseudonodosaria* sp. 1 | – | – | – | – | – | – | – | – | 1 | – | – | – | – | – | – | – |
| *Pullenia bulloides* | – | 15 | – | 1 | 6 | 2 | 2 | 3 | 4 | 2 | 1 | 6 | 6 | 2 | – | 2 |
| *Pullenia quinqueloba* | – | 3 | 2 | 1 | – | – | – | – | – | – | – | – | – | – | – | 1 |
| *Pullenia riveroi?* | – | – | – | – | – | – | – | – | – | – | – | – | – | – | – | – |
| *Pullenia subcarinata* | 1 | 2 | – | 1 | 2 | 3 | 1 | 4 | – | 1 | – | 2 | – | – | – | – |
| *Pyrgo depressa* | – | – | – | – | – | – | – | – | – | – | – | – | – | – | – | – |
| *Pyrgo murrhina* | 4 | 3 | 4 | 4 | – | 3 | 3 | 1 | – | – | – | 1 | 1 | 4 | 2 | – |
| *Pyrgo oblonga* | – | 3 | – | 1 | – | 1 | – | – | – | – | – | – | – | – | 1 | – |
| *Pyrulina extensa* | – | – | – | – | – | – | – | – | – | – | – | 2 | – | – | – | – |
| *Pyrulina fusiformis* | – | – | – | – | – | – | – | – | – | – | – | – | – | – | – | – |
| *Pyrulina gutta* | – | – | – | – | – | – | – | – | – | – | – | – | – | – | – | – |
| *Pyrulina* sp. 1 | – | – | – | – | – | – | – | – | – | – | – | – | – | – | – | – |
| *Pyrulina* sp. 2 | – | – | – | – | – | – | – | – | – | – | – | – | – | – | – | – |
| *Pyrulinoides* sp. 1 | – | – | – | – | – | – | 1 | – | – | – | – | 1 | 1 | – | – | – |
| *Pyrulinoides* sp. 2 | – | – | – | – | – | – | – | – | – | – | – | – | – | – | – | – |
| *Pyrulinoides* sp. 3 | – | – | – | – | – | – | – | – | – | – | – | – | – | – | – | – |
| *Quinqueloculina weaveri* | 2 | 4 | – | – | 1 | 1 | – | 2 | 3 | 1 | 2 | 3 | 1 | – | 3 | 2 |
| *Quinqueloculina venusta* | – | – | 1 | – | – | – | – | 1 | – | – | – | – | – | 2 | – | – |
| *Recurvoides scitulus* | – | – | 1 | 2 | – | – | – | – | – | – | – | – | – | – | 1 | – |
| *Rutherfordoides bradyi* | – | – | – | – | – | – | – | – | – | – | – | – | – | – | – | – |
| *Rutherfordoides tenuis* | – | – | – | – | – | – | – | – | – | – | – | – | – | – | – | – |
| *Saracenaria italica* | – | – | – | – | – | – | – | – | – | – | – | – | 1 | – | – | – |
| *Saracenaria latifrons* | – | – | – | – | – | – | – | – | – | – | – | – | – | – | – | – |
| *Sigmoilina tenuis* | – | – | – | 2 | 2 | 1 | – | 1 | – | – | – | – | – | – | – | – |
| *Sigmoilopsis schlumbergeri* | – | – | – | – | – | – | – | – | – | – | – | – | – | – | – | – |
| *Siphonodosaria abyssorum* | – | – | 1 | – | – | – | 9 | – | – | – | 3 | – | – | – | 15 | – |
| *Siphonodosaria consobrina* | – | – | – | – | – | – | 1 | – | – | – | – | – | – | – | 1 | – |
| *Siphonodosaria lepidula* | 1 | – | – | – | 2 | 2 | 1 | 3 | 1 | – | – | 4 | – | 2 | – | – |
| *Siphonodosaria simplex* | – | – | – | – | – | – | – | – | – | – | – | – | – | – | – | – |
| *Siphotextularia catenata* | 4 | – | – | 3 | 2 | – | – | 1 | – | 1 | – | – | – | – | – | – |
| *Stilostomella* sp. | – | – | – | – | – | – | – | – | – | – | – | – | – | – | – | – |
| *Triloculina tricarinata* | – | – | – | – | – | – | – | – | – | – | – | – | – | – | – | – |
| *Uvigerina auberiana* | 5 | – | – | – | 14 | 1 | 3 | – | – | – | – | 1 | 1 | – | 3 | 7 |
| *Uvigerina canariensis* | – | – | – | – | – | – | – | – | – | – | – | – | – | – | – | 1 |
| *Uvigerina hispida* | – | – | – | – | – | – | – | 3 | 1 | – | – | – | – | 3 | – | – |
| *Uvigerina hollicki* | – | – | – | – | – | – | – | – | – | – | – | – | – | – | – | – |
| *Uvigerina mantaensis* | – | – | – | – | – | – | – | – | – | – | – | – | – | – | – | – |
| *Uvigerina peregrina* | 5 | – | – | – | – | – | – | – | – | – | – | – | – | – | – | – |
| *Valvulineria humilis* | – | 1 | – | – | 1 | – | – | – | – | – | – | – | – | – | – | – |
| UNIDENTIFIED | – | – | – | – | – | – | – | – | – | 1 | – | – | 1 | – | 2 | – |

| SITE | | | | | | | | | | | | | | | | |
|---|---|---|---|---|---|---|---|---|---|---|---|---|---|---|---|---|
| CORE | | | | | | 18 | | | | | | | | 17 | | |
| SECTION | 3 | | 2 | | | 1 | | 3 | | | 2 | | | 1 | | |
| INTERVAL (cm) | 14-16 | 64-66 | 11-13 | 111-113 | 61-63 | 13-15 | 85-87 | 31-33 | 126-128 | 74-76 | 22-24 | 118-120 | 62-64 | 14-16 | 102-104 | 54-56 |
| DEPTH (mbsf) | 76.36 | 75.35 | 74.82 | 74.32 | 73.82 | 73.34 | 72.61 | 72.03 | 71.48 | 70.96 | 70.44 | 69.99 | 69.43 | 68.95 | 68.27 | 67.79 |
| AGE (MA) | 3.47 | 3.46 | 3.46 | 3.45 | 3.44 | 3.42 | 3.40 | 3.38 | 3.36 | 3.34 | 3.33 | 3.31 | 3.29 | 3.28 | 3.22 | 3.20 |
| *Pleurostomella recens* | 1 | – | – | – | – | – | – | – | – | – | – | – | 1 | – | – | – |
| *Pleurostomella subnodosa* | – | – | – | – | – | – | – | – | – | – | – | – | – | – | – | – |
| *Polymorphinidae formae fistulosae* | – | – | – | – | – | – | – | – | – | – | – | – | – | – | – | – |
| *Pseudonodosaria* sp. 1 | – | – | – | – | – | – | – | – | – | – | – | – | – | – | – | – |
| *Pullenia bulloides* | 1 | 4 | 8 | 1 | 1 | – | – | 2 | 3 | – | – | 5 | – | – | – | – |
| *Pullenia quinqueloba* | – | – | – | – | – | 6 | – | – | – | – | 4 | 3 | – | – | – | 3 |
| *Pullenia riveroi?* | – | – | – | – | – | – | – | – | – | – | – | – | – | – | – | – |
| *Pullenia subcarinata* | 2 | 1 | – | – | – | – | – | – | – | – | – | 1 | 2 | – | – | – |
| *Pyrgo depressa* | – | – | – | – | – | – | – | – | – | – | – | – | – | – | – | – |
| *Pyrgo murrhina* | 1 | – | 3 | 2 | – | 6 | 5 | 3 | 1 | – | – | 5 | – | 4 | 2 | 3 |
| *Pyrgo oblonga* | – | – | – | – | – | – | – | – | – | – | – | – | – | – | 2 | 1 |
| *Pyrulina extensa* | – | – | – | – | – | – | – | – | – | – | – | – | – | 1 | – | – |
| *Pyrulina fusiformis* | – | – | – | – | – | – | – | – | – | – | – | – | – | – | – | – |
| *Pyrulina gutta* | – | – | – | – | – | – | – | – | – | – | – | – | – | – | – | – |
| *Pyrulina* sp. 1 | – | – | – | – | – | – | – | – | – | – | – | – | – | – | – | – |
| *Pyrulina* sp. 2 | – | – | – | – | – | – | – | – | – | – | – | – | – | – | – | – |
| *Pyrulinoides* sp. 1 | – | – | – | – | – | – | – | – | – | – | – | – | – | – | – | – |
| *Pyrulinoides* sp. 2 | – | – | – | – | – | – | – | – | – | – | – | – | – | – | – | – |
| *Pyrulinoides* sp. 3 | – | – | – | – | – | – | – | – | – | – | – | – | – | – | – | – |
| *Quinqueloculina weaveri* | 2 | 1 | – | 6 | 3 | 5 | 5 | – | 4 | 4 | 4 | 4 | – | 2 | 1 | 2 |
| *Quinqueloculina venusta* | – | 1 | – | – | – | – | – | – | – | – | – | – | – | 2 | – | – |
| *Recurvoides scitulus* | – | – | – | – | – | – | – | – | – | – | – | – | – | 1 | – | – |
| *Rutherfordoides bradyi* | – | – | – | – | – | – | – | – | – | – | – | – | – | – | – | – |
| *Rutherfordoides tenuis* | – | – | – | – | – | – | – | – | – | – | – | – | – | 2 | – | – |
| *Saracenaria italica* | – | – | – | – | – | – | – | – | – | – | – | – | – | – | – | – |
| *Saracenaria latifrons* | – | – | – | – | – | – | – | – | – | – | – | – | – | – | – | – |
| *Sigmoilina tenuis* | – | – | – | – | – | – | – | – | – | – | – | – | – | – | – | – |
| *Sigmoilopsis schlumbergeri* | – | – | – | – | – | – | – | – | – | – | – | – | – | – | – | – |
| *Siphonodosaria abyssorum* | 7 | 1 | – | 1 | – | 1 | – | – | – | 1 | – | – | 1 | 1 | – | 1 |
| *Siphonodosaria consobrina* | – | 1 | – | – | – | – | – | 1 | 1 | 1 | – | – | – | 1 | – | – |
| *Siphonodosaria lepidula* | – | 1 | 3 | 2 | – | – | 1 | – | 3 | – | – | – | – | – | – | – |
| *Siphonodosaria simplex* | – | – | – | – | – | – | – | – | – | – | – | – | – | – | – | – |
| *Siphotextularia catenata* | – | 1 | – | – | – | – | – | – | – | – | – | – | – | 1 | 1 | – |
| *Stilostomella* sp. | – | – | – | – | – | – | – | – | – | – | – | – | – | – | – | – |
| *Triloculina tricarinata* | – | – | – | – | – | – | – | – | – | – | – | – | – | – | – | – |
| *Uvigerina auberiana* | – | – | – | – | 2 | 3 | – | – | – | 3 | 1 | 3 | – | 2 | 3 | – |
| *Uvigerina canariensis* | – | 1 | – | – | – | – | – | – | – | – | – | – | – | – | – | – |
| *Uvigerina hispida* | 1 | – | – | 2 | – | 4 | – | – | – | 3 | – | 2 | – | – | – | – |
| *Uvigerina hollicki* | – | – | – | – | – | – | – | – | – | – | – | – | – | – | – | – |
| *Uvigerina mantaensis* | – | – | – | – | – | – | – | – | – | – | – | – | – | – | – | – |
| *Uvigerina peregrina* | 4 | – | – | – | 1 | – | – | 3 | – | – | – | – | – | – | 4 | – |
| *Valvulineria humilis* | – | – | – | – | – | – | – | – | – | – | – | – | – | – | – | – |
| UNIDENTIFIED | 3 | – | – | – | – | – | – | – | – | – | – | – | – | 1 | – | 1 |

| SITE | | | | | | | | | | | | | | | | |
|---|---|---|---|---|---|---|---|---|---|---|---|---|---|---|---|---|
| CORE | | | | | 16 | | | 15 | | | | | 14 | | | |
| SECTION | 3 | 2 | | | 1 | | 3 | 2 | | 3 | 2 | | 1 | 3 | 2 | |
| INTERVAL (cm) | 1-3 | 100-102 | 146-148 | 96-98 | 44-46 | 61-63 | 11-13 | 61-63 | 77-79 | 28-30 | 128-130 | 128-130 | 78-80 | 37-39 | 81-84 | 126-128 |
| DEPTH (mbsf) | 67.26 | 66.75 | 65.87 | 65.37 | 64.85 | 63.27 | 62.77 | 61.77 | 59.00 | 58.51 | 58.00 | 56.89 | 56.39 | 54.53 | 53.47 | 52.47 |
| AGE (MA) | 3.19 | 3.17 | 3.14 | 3.13 | 3.11 | 3.07 | 3.03 | 2.99 | 2.89 | 2.87 | 2.84 | 2.79 | 2.78 | 2.69 | 2.63 | 2.61 |
| *Pleurostomella recens* | – | – | 1 | – | – | – | – | – | – | – | – | – | – | – | 1 | 1 |
| *Pleurostomella subnodosa* | – | – | – | – | – | – | – | – | – | – | – | – | – | – | – | – |
| *Polymorphinidae formae fistulosae* | – | – | – | – | – | – | – | – | – | – | 1 | 1 | – | – | – | – |
| *Pseudonodosaria* sp. 1 | – | – | – | – | – | – | – | – | – | – | – | – | – | – | – | – |
| *Pullenia bulloides* | – | 1 | – | – | 4 | – | – | 2 | – | – | – | – | – | 2 | – | – |
| *Pullenia quinqueloba* | 2 | – | – | – | – | – | – | – | – | 2 | – | – | 1 | 2 | – | – |
| *Pullenia riveroi?* | – | – | – | – | – | – | – | – | – | – | – | – | – | – | – | – |
| *Pullenia subcarinata* | 2 | – | – | – | – | – | – | – | – | 1 | 1 | – | – | – | – | – |
| *Pyrgo depressa* | – | – | – | 1 | – | – | – | – | – | – | – | – | – | – | – | – |
| *Pyrgo murrhina* | 5 | 1 | 2 | 3 | – | – | 3 | 1 | – | – | – | 2 | 2 | 1 | 3 | 2 |
| *Pyrgo oblonga* | 2 | – | – | 1 | – | – | – | – | – | – | 1 | – | – | – | – | 1 |
| *Pyrulina extensa* | – | – | – | – | – | – | – | – | – | – | – | – | – | – | – | – |
| *Pyrulina fusiformis* | – | – | – | – | – | – | 1 | 1 | – | – | – | – | – | – | – | – |
| *Pyrulina gutta* | – | – | – | – | – | – | – | – | – | – | – | – | – | – | – | – |
| *Pyrulina* sp. 1 | – | – | – | – | – | – | – | – | – | – | – | – | – | – | – | – |
| *Pyrulina* sp. 2 | – | – | – | – | – | – | – | – | – | – | – | – | – | – | – | – |
| *Pyrulinoides* sp. 1 | – | – | – | – | – | – | – | – | – | – | – | – | 2 | 1 | – | – |
| *Pyrulinoides* sp. 2 | – | – | – | – | – | – | – | – | – | – | – | – | – | 1 | – | – |
| *Pyrulinoides* sp. 3 | – | – | – | – | – | – | – | – | – | – | – | – | – | – | – | – |
| *Quinqueloculina weaveri* | 4 | 1 | 4 | 5 | 7 | 3 | – | 4 | 3 | – | 3 | 2 | – | 2 | 3 | 2 |
| *Quinqueloculina venusta* | – | – | 1 | – | – | – | 1 | 1 | – | – | – | – | – | – | – | 2 |
| *Recurvoides scitulus* | – | – | – | – | – | – | – | – | – | – | – | – | 1 | – | – | – |
| *Rutherfordoides bradyi* | – | – | – | – | – | – | – | – | – | – | – | – | – | – | – | – |
| *Rutherfordoides tenuis* | – | – | – | – | – | – | – | – | – | – | – | – | – | – | – | – |
| *Saracenaria italica* | – | – | – | – | – | – | – | – | – | – | – | – | – | – | – | – |
| *Saracenaria latifrons* | – | – | – | – | – | – | – | – | – | – | – | – | – | – | – | – |
| *Sigmoilina tenuis* | – | 1 | – | 1 | – | – | – | – | – | – | – | – | – | 1 | – | – |
| *Sigmoilopsis schlumbergeri* | – | – | – | – | – | – | – | – | – | – | – | – | – | – | – | – |
| *Siphonodosaria abyssorum* | – | – | – | – | – | – | 4 | – | 2 | 2 | 2 | – | 1 | – | – | – |
| *Siphonodosaria consobrina* | 3 | – | 1 | – | – | – | – | – | – | – | – | – | – | – | – | – |
| *Siphonodosaria lepidula* | – | 1 | 2 | 1 | 3 | – | 1 | – | 4 | 2 | – | – | 2 | 1 | – | 1 |
| *Siphonodosaria simplex* | – | – | – | – | – | – | – | – | – | – | – | – | – | – | – | – |
| *Siphotextularia catenata* | – | – | 5 | 2 | – | – | – | – | 2 | – | – | – | – | – | – | – |
| *Stilostomella* sp. | – | – | – | – | – | – | – | – | – | – | – | – | – | – | – | – |
| *Triloculina tricarinata* | – | – | – | – | – | – | – | – | – | – | – | – | – | – | – | – |
| *Uvigerina auberiana* | 4 | 3 | 4 | 6 | 1 | 3 | – | – | 4 | – | – | – | – | – | 2 | 1 |
| *Uvigerina canariensis* | – | – | – | – | – | – | – | – | – | – | – | – | – | – | – | – |
| *Uvigerina hispida* | 6 | 3 | 9 | 3 | 1 | – | – | – | 16 | 2 | – | – | 2 | 2 | – | – |
| *Uvigerina hollicki* | – | – | – | – | – | – | – | – | – | – | – | – | – | – | 1 | – |
| *Uvigerina mantaensis* | – | – | – | – | – | – | – | – | – | – | – | 1 | – | – | – | – |
| *Uvigerina peregrina* | – | – | – | – | – | – | 2 | 2 | – | – | 1 | – | – | – | – | 1 |
| *Valvulineria humilis* | – | – | – | – | – | – | – | – | – | – | 3 | 1 | – | – | – | – |
| UNIDENTIFIED | 1 | – | – | – | – | – | – | 2 | – | – | – | – | 1 | – | – | – |

| SITE | | | | | | | | | | | | | | | | 503B |
|---|---|---|---|---|---|---|---|---|---|---|---|---|---|---|---|---|
| CORE | | 13 | | | 12 | | | | 11 | | | | | 10 | 8 | 8 |
| SECTION | | 1 | 2 | | 1 | 3 | | | 2 | 3 | | 2 | | 1 | 3 | 1 |
| INTERVAL (cm) | 75-77 | 22-24 | 57-59 | 106-108 | 7-9 | 62-64 | 112-114 | 61-63 | 11-13 | 10-13 | 60-62 | 10-12 | 112-114 | 12-14 | 116-118 | 61-63 |
| DEPTH (mbsf) | 51.96 | 51.43 | 48.88 | 47.87 | 46.88 | 45.80 | 44.80 | 44.29 | 43.79 | 41.08 | 40.08 | 39.58 | 39.13 | 38.13 | 33.28 | 29.82 |
| AGE (MA) | 2.59 | 2.57 | 2.46 | 2.42 | 2.38 | 2.33 | 2.29 | 2.27 | 2.25 | 2.14 | 2.09 | 2.07 | 2.05 | 2.01 | 1.81 | 1.66 |
| *Pleurostomella recens* | – | – | – | – | – | – | – | 1 | – | – | – | 1 | – | – | – | – |
| *Pleurostomella subnodosa* | – | – | – | – | – | – | – | – | – | 1 | – | – | – | – | – | – |
| *Polymorphinidae formae fistulosae* | – | – | – | – | – | – | – | – | – | – | – | – | – | – | – | – |
| *Pseudonodosaria* sp. 1 | – | – | – | – | – | – | – | – | – | – | 1 | – | – | – | – | – |
| *Pullenia bulloides* | 2 | – | 4 | 2 | 1 | – | 1 | 1 | 3 | 1 | – | 3 | 10 | 1 | 1 | 1 |
| *Pullenia quinqueloba* | 1 | – | – | – | – | – | 1 | 1 | – | – | – | 1 | – | – | – | – |
| *Pullenia riveroi?* | – | – | – | – | – | – | – | – | – | – | – | – | – | – | – | – |
| *Pullenia subcarinata* | – | – | 1 | – | – | – | – | – | – | – | – | 1 | 3 | – | – | – |
| *Pyrgo depressa* | – | – | – | – | – | – | – | – | – | – | – | – | – | – | – | – |
| *Pyrgo murrhina* | – | 1 | 1 | 1 | – | 1 | 2 | – | – | 2 | 1 | – | 3 | 4 | 4 | – |
| *Pyrgo oblonga* | – | 2 | 1 | – | – | – | 2 | – | – | – | – | – | 2 | 2 | – | – |
| *Pyrulina extensa* | – | – | – | – | – | – | – | – | – | – | – | – | – | – | – | – |
| *Pyrulina fusiformis* | – | – | – | – | – | – | – | – | – | – | – | – | – | – | – | – |
| *Pyrulina gutta* | – | – | – | – | – | – | – | – | – | – | – | – | – | – | – | – |
| *Pyrulina* sp. 1 | – | – | – | – | – | – | – | – | – | – | – | – | – | – | – | – |
| *Pyrulina* sp. 2 | – | – | – | – | – | – | – | – | – | – | – | – | – | – | – | – |
| *Pyrulinoides* sp. 1 | 1 | – | – | – | – | – | – | – | – | – | – | – | – | – | – | – |
| *Pyrulinoides* sp. 2 | – | – | – | – | – | – | – | – | – | – | – | – | – | – | – | – |
| *Pyrulinoides* sp. 3 | – | – | – | – | – | – | – | – | – | – | – | – | – | – | – | – |
| *Quinqueloculina weaveri* | 6 | 3 | 4 | – | – | 2 | 1 | 2 | 1 | 2 | 2 | 3 | 1 | 1 | 5 | 1 |
| *Quinqueloculina venusta* | 1 | – | 1 | – | – | – | 2 | – | 1 | – | – | – | – | 1 | – | – |
| *Recurvoides scitulus* | – | – | – | – | – | – | – | – | – | – | – | – | – | – | – | – |
| *Rutherfordoides bradyi* | – | – | – | – | – | – | – | – | – | – | – | – | – | – | – | – |
| *Rutherfordoides tenuis* | – | – | – | – | – | – | – | – | – | – | – | – | – | – | – | – |
| *Saracenaria italica* | – | – | 1 | – | – | 1 | – | – | – | – | – | – | – | – | – | – |
| *Saracenaria latifrons* | – | – | – | – | – | – | – | – | – | – | – | – | – | – | 1 | – |
| *Sigmoilina tenuis* | – | – | – | 1 | – | – | – | – | – | – | – | – | – | – | – | – |
| *Sigmoilopsis schlumbergeri* | – | – | – | – | – | – | – | – | – | – | – | – | – | – | – | – |
| *Siphonodosaria abyssorum* | – | – | – | – | – | – | – | – | – | – | – | – | – | – | – | – |
| *Siphonodosaria consobrina* | – | – | – | – | – | – | – | – | – | – | – | – | – | – | 1 | – |
| *Siphonodosaria lepidula* | – | – | – | – | – | – | – | – | – | – | – | – | – | – | – | – |
| *Siphonodosaria simplex* | – | – | – | – | – | – | – | – | – | – | – | – | – | – | – | – |
| *Siphotextularia catenata* | 3 | – | – | 1 | – | – | 1 | – | – | – | – | – | – | – | – | – |
| *Stilostomella* sp. | – | – | – | – | – | – | – | – | – | – | – | – | – | – | – | – |
| *Triloculina tricarinata* | – | – | – | – | – | – | – | – | – | – | – | – | – | – | – | – |
| *Uvigerina auberiana* | 2 | – | 1 | – | – | – | – | – | – | 2 | – | – | – | – | – | – |
| *Uvigerina canariensis* | – | – | – | – | – | – | – | – | – | – | – | – | – | – | – | – |
| *Uvigerina hispida* | – | – | – | – | – | – | – | – | – | 7 | – | – | – | 3 | – | – |
| *Uvigerina hollicki* | – | – | – | – | – | – | – | – | – | – | – | – | – | – | – | – |
| *Uvigerina mantaensis* | – | – | – | – | – | – | – | – | – | – | – | – | – | – | 1 | – |
| *Uvigerina peregrina* | – | – | – | – | 1 | – | – | – | – | – | – | – | – | – | – | 3 |
| *Valvulineria humilis* | – | – | – | – | – | – | – | – | – | – | – | – | 1 | – | – | – |
| UNIDENTIFIED | – | – | – | – | – | – | – | – | – | – | – | – | – | – | – | – |